About the Author

Daisy Kendrick interned at the United Nations in New York for the Permanent Mission of Grenada after graduating from Northeastern University in Boston. Following this she founded Ocean Generation (OG) to disrupt and innovate the standard charity model. Frustrated at the lack of awareness and action being taken by millennials and Gen-Z to protect our climate and oceans, she created OG to utilise media and technology to inform, educate and change behaviour at a global scale. She is the youngest recipient of the 2018 *Marie Claire* Future Shapers award and was recognised as one of Britain's 50 most remarkable women by the *Daily Mail*.

DAISY KENDRICK
FOUNDER OF OCEAN GENERATION

THE CLIMATE IS CHANGING

WHY AREN'T WE?

For every book purchased, the author will donate proceeds to plant trees around the world.

PIATKUS

First published in Great Britain in 2020 by Piatkus

1 3 5 7 9 10 8 6 4 2

A CIP catalogue record for this book is available from the British Library.

ISBN 978-0-349-42392-0

Typeset in Avenir by M Rules
Printed and bound in Great Britain by Clays Ltd, Elcograf S.p.A.

Papers used by Piatkus are from well-managed forests and other responsible sources.

Piatkus
An imprint of
Little, Brown Book Group
Carmelite House
50 Victoria Embankment
London EC4Y 0DZ

An Hachette UK Company
www.hachette.co.uk

Contents

Acknowledgements

Thank you to those who have shared their stories, thoughts, quotes and words – they have helped put this book together to create a collective of voices for the climate narrative.

My thanks also go to:

Brad Davidson for being a constant soundboard and support from the very beginning of this book journey. Your ideas and input were incredibly valuable.

My publisher, Jillian Young, for granting me this opportunity and being so patient along the way.

To Little, Brown Book Group for recognising that climate change is a real threat and that the more literature we can have on this topic the more we can spread awareness of the need for change.

This book is a voice for those people whose lives are dramatically changing because of climate change. The areas first and most affected are the poorest regions of the Earth; they are also the least represented on our planet and don't have the platform to share their stories. This book is for those who aren't even here; they haven't been born yet. It asks you to feel empathy for the unborn and the poor of the global south and to change your lifestyle based on their needs – something that humanity has never done before.

Introduction

'Do the best you can until you know better. Then, when you know better, do better'

Maya Angelou

You might be reading this, sitting in a trendy coffee shop with a steaming oat latte (oat, because dairy is frowned upon in those trendy places). On the table next to you is your iPhone, where you have just flicked through your Instagram account, a social media platform filling your subconscious with potentially pointless but often beautiful-looking content. Right about now, a waiter is walking towards you with your avocado on toast.

Maybe you should snap it and share the picture with friends? A great use of time. Or would it be more useful to continue reading this book and learn that it takes on average 272 litres of water to grow a single avocado? That one little fruit has its own carbon footprint, and it might even have contributed to a devastating drought in the region where it grew. (Don't worry if you're not the basic brunching-avocado-and-toast kind, there will be lots more examples as you read on.)

How much hotter will our planet get? When will we see the planet's temperature rise by two, three or even four degrees Celsius? How much sea-level rise will be here in 2030, 2050 and 2100? Which cities will flood first, which forests will dry out, and which populations will die of hunger first?

Excuse me – what?

Through human activity we have managed to grow the monster that is climate change. The greatest challenge humanity is facing. The challenge is perhaps beyond imagination, and the consequences are far more involved with politics, economics and self-interests than any of us might care to believe. It is also most certainly easier to ignore climate change if you can't see it where you live and therefore it is a greater excuse for inaction. Or, if you are simply on the climate denier side, the prospect of climate change hasn't been keeping you awake at night. This book might just change your mind.

From London to New York to Mumbai, many people are only dimly aware, if at all, of the severity of climate change. As global temperatures rise, they will cause a myriad of consequences such as natural disasters, famine and droughts. We know that humans have wiped out 60 per cent of all animal populations in just 40 years.[1] Our oceans are so polluted with plastics that if you eat fish you are most likely digesting microplastics as a result of fish mistaking plastics for food. Climate change is today confronting humanity with the hardest trials we have ever encountered, and it will continue to do so. Climate change might soon push billions of people out of their homes, countries and jobs, and create more upheaval than any existing ideology knows how to handle.

Shit! Pretty scary times.

The intention of this book, though, is not actually to make you feel scared. Despair leads to further inaction and a future that Sir David Attenborough has described as 'the collapse of our civilisations and the extinction of much of the natural world'. Optimism and engagement, on the other hand, can help us to avoid further catastrophe for people and the planet.

We are in an era of urgent calls to action, but why do most of us ignore this urgency? The science of global warming has been settled for 40 years, but instead of listening to the warnings we have continued to pollute at an accelerated rate. Governments, scientists and corporations know what is happening; they have known for years. Governments and scientists of conscience need to break or redefine their ties with corporations or individuals financing the injustices of climate change globally. For far too long humanity thought it could selfishly draw from infinite resources of the Earth without consequences, but the world is not a commodity. If we take civil and human rights seriously, we cannot assign solving the global climate crises solely to the individual; we need to realise the scale of political and commercial change that needs to happen.

It is true that we don't actually know precisely how it will all play out, although irreversible changes to the planet are already underway, but in order for us to never find out what World War III or an apocalypse could look like, you, the reader, must start acting.

Throughout this book I explore in-depth the reality of climate change and how different industries are affecting the planet, and provide ideas – and easy switches – so that you, as an individual, can play a vital role in being a catalyst for change for future generations.

Ideally, every reader would make every switch I suggest, but that's not realistic. This book is not about you switching to veganism today and becoming zero waste right now to save our oceans; this book is about the small collective actions. What you take from this book has to work for your lifestyle, your economic situation and, ultimately, what makes you happy and what you believe in.

If you are questioning how willing you are, or why you should deprive yourself of brief pleasures, such as eating meat every day or taking long, hot showers, that is completely natural, but it ignores the realities we are faced with. And if you think that this is the fault of the West and only a problem for the rich – yes, darling, ski season in Verbier *will* be affected without a doubt by inconsistent snowfall – be aware that almost everyone reading this is living a better life than 99 per cent of all human beings who have ever lived. We must exercise our privilege to do more.

Opting for less consumption or healthier choices in no way means living a miserable life; on the contrary, it can mean living a more meaningful life.[2] Politics, legislation and huge corporations are largely to blame when it comes to climate change, and they need to get their own shit together without further hesitation. The scientists, global leaders and businesses have continued to feed our polluting consumption habits. If money, short-term attention spans and fossil-fuel interests were not the current vision of politics, but all life – from people to biodiversity – were at the centre, we would be in a much better place.

Until we can organise politics and business around a vision that demands a safer future, the only people we can truly rely on are ourselves. If they won't change, and politicians are asking the individual to change, it has really come to a global tipping point. We can no longer rely on others to help people and the planet – it really is up to us.

Yes, that means *you*.

It is important to be honest about what individual eco-switches can and will do immediately for the greater picture of climate change. If every consumer immediately stopped using cars and planes, buying fast fashion, eating meat and using plastic, the problems would not, in fact, be fixed overnight. These efforts are but a drop in the ocean of the magnitude of damage humans have caused our planet in the last 100 years. However, highlighting the truths about how many of us live, and the world that we as consumers have built, in a powerful and candid way, will give you the facts about various industries and provide switches for every reader at different stages of their life.

A not-yet-perfect author

My climate journey was not straightforward. I was not born an environmentalist and my upbringing never conditioned me to think outside my societal bubble, but that all changed in 2015 when I did an internship at the United Nations (UN) in New York. It was an incredibly exciting year, when 194 countries negotiated and signed into effect the Sustainable Development Agenda, an urgent call to action by all countries, developed and developing, in a global partnership of peace for people and the planet.

For the first time in history, climate change was high on the agenda with goals dedicated to nature-based themes. My long intern days were spent running around from meeting to meeting listening and taking notes on the rapid death of our oceans, loss of biodiversity, air pollution and ultimately human suffering of millions of people globally because of climate change.

Disillusioned and overwhelmed by the complexity of climate change, I began getting involved with local organisations and eventually founded, at the age of 21, my own non-profit called the Ocean Generation Foundation. There are thousands of environmental non-profits globally, so it wasn't vital that I started my own, but I felt that there was a huge gap in the space for relatable and cool content that could connect millennials and Gen-Z with environmentalism on their terms.

By experimenting with music, mobile gaming and coding, Ocean Generation was trail-blazing a new way of communicating climate change to the most connected generation. We reached millions of young people across a magnitude of platforms, planting the seed of what was happening around the world in terms of climate issues and ideas that could be implemented both on an individual and larger scale for a safer planetary future. From awareness, we began expanding into grass-roots action with initiatives empowering hundreds of young people from climate-vulnerable communities on islands with new skills and using technology to come up with solutions to different ocean and climate-specific issues.

As I will continue to emphasise throughout this book, climate change is a complex topic. It is very science based, the statistics are incredibly negative, and we often forget the social aspects of climate change: the people who are dying, losing homes, getting ill and struggling to eat. The topic often seems unrelatable to many of this generation, but through every loss I have witnessed while working with communities and travelling the world, I also see glimmers of hope, because nature is so resilient. Nature works in cycles, and sometimes, miraculously, things can come back to life, and therefore we should change our habits and never lose hope that one small action, multiplied by millions of people, can have a positive impact.

Starting a climate organisation has allowed me to see both sides of the climate conversation. I am now part of high-level discussions on how we can solve one of humanity's greatest challenges, but it is important to highlight that I was never brought up in an environmental background. I had absolutely no clue that climate change was really happening. The extent of my family or my social circle doing something 'eco' was, at most, using a recycling bin. A greener life was not induced in me from a young age and still today I am no expert. I openly acknowledge whatever is said in this book will never be enough for some people, and a myriad of backlash and criticism might follow.

I occasionally wear leather and I go on a *lot* of aeroplanes because of my job. I do not live a 100 per cent sustainable life, but I also live in a major city, which makes it incredibly difficult. In many towns and cities, just like London, across the world, interacting with plastic or using fossil fuels is simply unavoidable (almost everything we use is here because of fossil fuels), and we have been conditioned as societies to just go with it and not be uncomfortable challenging the norms. I am also in a position of privilege because I have time and money to think about making alternative decisions, such as where to shop or how to eat to align with my ethics, which is not a freedom everyone has. The important thing is to recognise the bigger picture through being educated about what is happening around the world and making conscious decisions about individual lifestyle choices with those in mind.

When I was growing up I had no idea that by the time I became a young woman our planet Earth would be in total jeopardy for my children. I was too busy living, too busy consuming, too busy falling victim to societal norms. Then I realised that our lives play a vital role in the lives of others, too, and it is all our individual responsibilities to be kinder and make the world a better place.

Since learning about the environment, I have taken many moments to reflect and learn. Constant searching has led me to be fascinated by the world of nature and all its beauty, and it's something I wanted to emphasise in this book. Without nature, nothing and no one can exist. Nature is everything and it is what keeps us alive.

After years of philanthropic work and travelling the world talking to people dealing with various levels of climate catastrophe,

I decided that I am tired of the stigma climate change activism entails. I am tired of showing up to events and seeing the same people, tired of watching huge brands greenwashing (I'll explain about that term in Chapter 1) and tired of the 'one way or no way' approach to solving climate change. Activism is not an exclusive club; we are all activists in our own right, and together we become a global collective created to safeguard the future of our planet.

We are living in an incredible technologically advanced world where innovations are popping up everywhere trying to rewrite the wrongs of the past, and start-up businesses have purpose at their core. Every single action we take has an impact on the environment. Each one of us has something unique to offer back to the world, which only we can give based on our circumstances, and for that reason no one solution that becomes a trend across media will ever solve climate change. Let us humanise the climate change story and use our collective responsibility as individuals, brands, entrepreneurs, governments and citizens to get our shit together.

Like every other human being I am still learning, and I am on a journey learning how to be a more environmentally friendly young woman living in a cosmopolitan city. I am no better than anyone else, but some of the information and tips that you will find in this book have helped me to live a more conscious life.

The lessons in this book are not rules, nor will they ever become the status quo. Some switches in the book might sound like common sense, or you might have seen them hundreds of times across social media. The goal is to simply talk about different environmental impacts that various industries are having and to interpret ways in which we as individuals can make easy changes in our lives.

Why this book?

This book aims to serve as an inspiration and tool that focuses on individual change for climate action.

The severity of climate change and its potential impact on our livelihoods should leave nobody indifferent. This book will open your mind with facts about different industries, from the clothes you buy to the food you eat, to what you do at home, and everything in between. From the plastics we use that fuel climate change and jeopardise our future, to our use of social media and technology, the surge of modern-day activism and what tools are empowering us to get our voices heard and demand climate action.

Each chapter first delves into how each industry affects the environment, the blows it has on people and communities globally, and, finally, it suggests simple switches that you can make in your everyday life. Making a difference and doing your part for the environment doesn't mean taking up all your time and money. Some of the eco-switches might seem obvious, some may be unattainable to you, and some might not be even relevant.

The aim of all of this is to empower you with knowledge so that you can make decisions that fit your beliefs and lifestyle. It will give you practical ways through which you, as a consumer, can reduce your own impact in this deeply complicated problem, and you will know that small changes multiplied by millions of individuals will change how we consume, what brands and governments start to prioritise and what we pass on to future generations.

The latter part of the book touches upon how social media, business and technology influences our views on climate change and what those companies can do collectively to spread the climate change message positively. We will also look at how women are affected by climate change and how it affects their day-to-day lives. The term 'eco-feminism' is discussed along with the under-representation of female voices in climate negotiations and policy creation, and the necessity for more inclusion. Lastly, I talk about people and the impact of climate change on communities around the globe. The people aspect is dear to my heart. It is the side of climate change that the mainstream media tends to neglect in favour of cute sea creatures, but the reality is that when a disaster strikes it takes decades to rebuild communities and we, as humankind, need to be ready to adapt to our rapidly changing circumstances.

When making any of the switches suggested in this book it becomes a lot more powerful to think that this is not just for yourself but for *all* humankind, including those not yet born.

Throughout the book incredible activists and change-makers in the environmental space are highlighted from all around the world, and, wherever possible, they have contributed in their own words. This climate journey is not mine to tell, and it is a privilege to hear from others sharing their stories and work. Combined, this book becomes a powerful collective.

When reading information about our world, it is not my intention to send you as a reader into any type of hysteria, but instead to switch those emotions from panic mode to action mode. The first step in making a better world is caring about things that might not directly affect you right now, and, amazingly, your first step is reading this book. It is down to collective action to save all of us, and future generations. It is the importance of realising that every time you swipe your card and make a purchase, it has a knock-on effect somewhere else in the world and will directly or indirectly contribute to climate change.

The intention of this book is not to tell you what you can and can't do or achieve if you care about our planet, nor how to live your life, neither does it claim that only certain ways of living are correct. The facts, stories and easy green switches are to enable you to make more informed life choices and grant future generations the rights they deserve.

The climate is changing, so why aren't we? After all, we are the generation of change.

We Are

by Nils Leonard

Some are happy to sit and watch.
To settle.
To carry on the way things are.

And some see it differently.
They see the way things really are.
They want to make a change.

To them the world isn't distant.
They see it in everything they do.

In how they eat, drink, work, love and live.
They understand that the Earth is in everything.
They understand the power
In every moment to change the future of our planet.

This isn't about money. It's about us. All of us.

And it is time for us,
To vote with our lives,
Our daily lives,
For or against the vision of a more hopeful future.

We see a new way.
We will change the world's relationship with plastic.

We will support those that rely on the oceans for life.
We will innovate, partnering those bringing new life to the planet.

We are powerful.
We are change.
We are the workers.
The artists. The thinkers.
The most connected generation ever to walk the planet.

We are ready.
We are one.

We are the global collective bringing climate change and its importance to the most connected generation in history.

1

Climate Change: The Basics

'Climate change is no longer some far-off problem; it is happening here, it is happening now'

Barack Obama

First of all let's get some definitions straight. Labels such as 'global warming', 'climate change', 'sustainability' and 'organic' have become corrupted and meaningless because of their misuse, and this is beginning to obscure the real issues that must be dealt with. Environmentalism is a hot topic, but its associated words are frequently inaccurately overused, and even used as clickbait for millions of online daily articles. Don't be fooled by what you read on the Internet, because climate change and global warming are *not* the same thing: one is a result of the other.

An eco dictionary

This little eco dictionary lists some of the terminology used in this book and regularly in climate discussions. I'll give my apologies upfront, because, after hours trawling through Google on how to make any of the climate language sound more fun, there really is no more enjoyable way to explain these terms without using their very sciency jargon; however, they are all words you should know, considering the state of our planet.

Take a screen shot of these words using your phone in case you ever find yourself in the midst of a climate conversation and need to sound smart.

Adaptation is an action that helps to cope with the effects of climate change; for example, the construction of barriers to protect against rising sea levels, or farmers changing to crops capable of surviving higher temperatures and drought.[1]

Carbon footprint This is the amount of carbon dioxide (CO2) and greenhouse gases released into the atmosphere as a result of your individual activities. Consider everything you do in life to calculate your footprint, from the food you eat, your electric usage, aeroplane trips, car journeys, playing sport under floodlights, and the clothes you buy. (A study by the Carbon Trust puts the annual carbon footprint of the average Brit at 10.92 tonnes of CO2.)

Carbon offsetting A way of compensating for emissions of CO2 by participating in, or funding, efforts to take CO2 out of the atmosphere. Offsetting often involves paying another party, somewhere else, to save emissions equivalent to those produced by your activity.[2]

Circular economy is an alternative to a traditional linear economy (based on make, use, dispose) in which we keep resources in use for as long as possible, extract the maximum value from them while in use, then recover and regenerate products and materials at the end of each service life.[3]

Climate change is any significant long-term change in the expected patterns of average weather of a region (or the whole Earth) over a significant period of time.

Climate resilience is the ability of an ecosystem to absorb, withstand and bounce back after an adverse event.

Deforestation The cutting down of trees in a large area, or the destruction of forests by people.[4]

Fossil fuels are natural resources, such as coal, oil and natural gas, which contain hydrocarbons. These fuels are formed in the Earth over millions of years and produce CO2 when burnt.

Global warming is the long-term average temperature increase of the Earth's climate system, produced by adding greenhouse gases (see overleaf) to the atmosphere.

Greenhouse gases (GHGs) The main greenhouse gases, such as CO2 and methane, have different global-warming potential. Greenhouse gases combined with water vapour act like a blanket, keeping in some of the sun's warmth. Increasing the amount of GHGs through burning fossil fuels is like wrapping the Earth in a thicker blanket. This increase is called 'global warming'.[5]

Greenwashing is the practice of making an unsubstantiated or misleading claim about the environmental benefits of a product, service, technology or company practice. Greenwashing can make a company appear to be more environmentally friendly than it really is.

IPCC The Intergovernmental Panel on Climate Change, established by the United Nations Environment Programme and the World Meteorological Organization, is a body for assessing the science relating to climate change. It is dedicated to providing the world with an objective, scientific view of climate change, its natural, political and economic impacts and risks, and possible response options. The science agreed by the IPPC is undisputed and non-negotiable: the climate and ecological crises pose an imminent threat to life on Earth.

The rapid pace of science and new discoveries are constantly changing within the climate field and there will be so many other topics, so to speak, that come under the umbrella of climate change in the near future.

Mitigation Action that will reduce human-made climate change. This includes action to reduce greenhouse gas emissions or absorb greenhouse gases in the atmosphere.[6]

Ocean acidification is a term used to describe significant changes to the chemistry of the ocean. It occurs when carbon dioxide is absorbed by the ocean and reacts with seawater to produce acid.[7] Although it occurs naturally, the higher quantities of CO2 in the atmosphere have caused higher levels of acidification. The consequences of this include inhibited coral growth and disorders in some sea life.

Organic farming is the name given to a variety of agricultural production systems that aim to work with nature rather than by conquering it.

Renewable energy is energy that can be replenished in a short period of time. The five most common renewable sources are: solar, wind, water, geothermal and biomass. By reducing carbon dioxide from energy, a decarbonised economy is an achievable objective by governments.

Sustainable development is the kind of development that satisfies the needs of the present without adversely affecting the ability of future generations to satisfy their needs.[8]

Sustainable development goals (SDGs) are a collection of 17 global goals set by the United Nations General Assembly in 2015 for the year 2030. The goals and targets are universal, meaning that they apply to all countries in the world, with a vision of ending poverty, protecting the planet and ensuring that all people enjoy peace and prosperity.

Note: while you're reading, if other environmental words come to mind, or if you come across words in the future that you would like to take a note of, feel free to use the space on page 259 at the back of the book to jot them down.

The sixth mass extinction

For over 3.5 billion years, living organisms have thrived, reproduced and diversified to dwell in every ecosystem on Earth. On the flip side of the burst of new species, species extinctions have also always been part of the evolutionary life cycle. But when these two processes are not aligned, and the loss of species outpaces the birth of new ones, this can tip us into what we know as a 'mass extinction' event, defined as a loss of about three quarters of all species on Earth over a 'short' geological period of time. I say 'short' because given the vast amount of time since life first evolved on the planet, 'short' is defined as anything less than 2.8 million years.

Since at least the Cambrian period, which began around 540 million years ago, when life first exploded into an array of forms, only five extinction events have met the mass-extinction criteria according to science. These so-called 'Big Five' have become part of a scientific benchmark to determine whether humans today have created the conditions for a sixth mass extinction.[9] Each of the extinctions has happened on average every 100 million years or so since Cambrian, and each event lasted between 50 thousand and 2.76 million years.

The Big Five

Ordovician period The first mass extinction happened around 444 million years ago and wiped out over 85 per cent of all species. The event seems to be the result of two climate phenomena: a global-scale ice age followed by a rapid warming period.

Late Devonian period Around 375 million years ago Earth was characterised by high variation in sea levels, and rapidly alternating conditions of global warming and cooling. It was also a time when plants were starting to take over dry land and there were drops in global CO2 concentrations. This affected 75 per cent of all species, most of which were bottom-dwelling invertebrates in tropical seas at the time.

Permian period Perhaps the most extreme and devastating mass extinction to occur was around 250 million years ago, when more than 95 per cent of all species in existence were wiped out. Some of the possible causes are still debated, such as massive volcanic activity in what is now Siberia, or increasing ocean toxicity caused by an increase in atmospheric CO2, or an asteroid impact that filled the air with pulverised particles that could have blocked the sun and caused intense acid rains. All of these changing conditions created unfavourable conditions for too many species.

Triassic-Jurassic period Fifty million years after the great Permian extinction, another 80 per cent of the world's species became extinct during the Triassic event. As a result, again, of elevated CO2 levels, some colossal geological activity in what is today the Atlantic Ocean and increased global temperatures and acidified oceans.

Cretaceous period The last and most famous mass extinction happened 66 million years ago, when an estimated 76 per cent of species became extinct, including dinosaurs. The demise of the dinosaur appears to have given mammals, from which human beings eventually evolved, the opportunity to occupy new habitats.

The Sixth Extinction: Today? The planet is currently experiencing a biodiversity crisis due to the exploitation of the Earth by people. Recent estimates suggest the extinction threatens up to a million species, and is driven by a mix of direct and indirect human activity, such as the destruction of habitats, chemical pollution, exploitation of fishing and hunting and burning fossil fuels. Today, extinctions are occurring hundreds of times faster than they would naturally, and if this continues at its current rate, we could approach the level of mass extinction as soon as

240–540 years. As previous extinctions demonstrate, climate change can be profoundly disruptive, and only a few generations down the line people may have to deal with the consequences and witness the damage that would throw ecosystems into chaos, reshaping our world.

Let's be clear, climate change did not suddenly start becoming a 'thing' with the rise of social media and the endless content on plastic pollution. We have scientific recollection about historical mass extinctions, scientific journals dating back to the 1890s and computer models of global climate since the 1960s. But only in the last 40 years has climate change become newsworthy. A century of accumulated science has taken time to build a budding case against fossil fuels, endangered rain forests, acidic oceans and a plethora of other consequences of climate change.

A pioneering Swedish scientist called Svante Arrhenius first estimated the scope of global warming from widespread coal burning in 1896. There were further historical perspectives on climate change and scattered reports throughout the decades. Then, in 1956, an article in the *New York Times* conveyed the long-lasting environmental changes that would ensue from the accumulation of greenhouse gas emissions from energy production. The article, however, foresaw that if coal and oil were plentiful and cheap globally, there was every reason to believe that both would be consumed if it paid producers to do so.[10]

Consequently, in 1988, a variety of factors, including drought and fires, peaked around the world and thrust the greenhouse effect into the spotlight. That same year the IPCC was established. Ideas to eliminate certain compounds threatening the atmosphere's protective ozone layer came into play under the Montreal Protocol agreed by nations worldwide.

The Montreal Protocol was the most successful environmental protection agreement and the first international climate treaty to achieve complete ratification. It set out binding progressive obligations for countries to phase out 96 ozone-depleting chemicals in thousands of applications across more than 240 industrial sectors. This was all within a stable framework that allowed industries to plan long-term research and innovation. To their credit, chemical companies have kept innovating and are now producing chemicals with no ozone-depleting potential.

In retrospect this was also the time to change and develop alternatives to minimise the cost of the damage to ecosystems by fossil fuels, but, instead, leaders and corporations turned a blind eye. There was also a lack of emphasis on renewable energies. In the 1970s, environmentalists promoted the development of renewable energy, both as a substitute for the eventual replacement of oil and as an escape from our dependence. The use of renewable energy sources and the rational use of energy are fundamental inputs to any responsible energy policy.

The conversation around nuclear energy wasn't highlighted enough in the Montreal Protocol: while nuclear doesn't emit greenhouse gases, it is incredibly dangerous, expensive and creates mountains of toxic radioactive waste. Energy and fossil fuel conversations only became more focused years later in other protocols, when 'climate change' became a more prominent topic for conversation.

The threat to climate systems

Scientists have concluded that human-generated emissions such as carbon dioxide and methane cause global warming, and so does deforestation. This global warming threatens climate systems, producing hotter temperatures on land and at sea, and this changes how and where precipitation (rain) falls. Those shifting weather patterns exacerbate catastrophes such as droughts, floods, wildfires, cyclones, loss of biodiversity and hurricanes. They also melt ice caps and glaciers, resulting in rising sea levels and threatening entire small-island nations.

Changing temperatures sway entire ecosystems, too, pushing migration patterns and food production in some regions completely out of whack. This is why we have started to see climate change endangering human health. As climate change fuels temperature increases and extreme weather events, it jeopardises our vital resources such as reducing the air quality, the supply of water and food, spreading disease and endangering our homes.[11] We are reaching a tipping point of global climate catastrophe beyond which Earth will become unbearable for all life.

Pollution is degrading the quality of elements around us, making our environment unsafe and unsuitable. Air pollution is considered a silent killer, especially for those who live in cities.

Not only are these people more likely to develop lung cancer as a result, but also every year over 7 million people are victims of polluted air-related deaths. Changing the average temperature of an entire planet, even if it's just by a few degrees, is a big deal. Right now the world is 1.2°C warmer than it was in pre-industrial times because of more CO2 and burning fossil fuels. At a 2°C temperature rise, air pollution will create an additional 153 million deaths, because temperature inversions change the dynamics of air movement. Warm air rises in the atmosphere because it is less dense and more buoyant than the cooler air above it. This smothering effect traps pollutants and allows their concentrations to increase. Essentially the heat and sunlight cook the air along with all the chemical compounds lingering within it, creating a smog horrible to inhale, especially for vulnerable groups such as the young and elderly. If we continue the current trajectory, climate change comes close to wiping out the last 50 years of gains we have seen in public health. Air pollution also causes damage to crops, animals and water, because it contributes to the depletion of the ozone layer, which protects the Earth from the sun's UV rays. In areas with high air pollution, agriculturalists are seeing severe reductions in crop yields, as well as nutrient limitations, which is not helpful – we need more food to feed our growing population.

The rapid deforestation of forests and jungles caused by burning to make land available for animals is alarming and contributes to 15 per cent of greenhouse gas emissions.

Plants play a vital role in our lives and our planet – in fact we need them in order to survive – so it is important to prevent illegal logging and the depletion of greenery, for example in the Amazon. Our oceans, too, are in danger of pollution due to the plastic contained in them, oil spills and other pollutants, and drinking water supplies are also under threat. More than half the world's corals have died due to climate change – more specifically as a result of coral bleaching. When corals experience stress from increased water temperatures or poor water quality, coral ejects a photosynthetic algae, which removes the coral's distinctive colours. If the stress conditions persist, the coral will die, but if conditions return to acceptable levels, some coral can reabsorb the substance and recover. Recovery is contingent on environmental conditions, so there is still hope.[12]

Several regions in the world are suffering from a reduction in water quality and a shortage of water supply. The decline in the quality of drinking water leads to higher bottled water consumption, but in regions with fewer resources, bottled water might not be a possibility, resulting in dehydration, sickness and even death.

Definitions are evolving and details change, and many more terms have come under the umbrella of climate change, but the main issues remain the same as they were 40 years ago. It is uncomfortable to think of the missed opportunities that let the problem worsen, and we now have to talk about climate change due to global warming, which is much harder than trying to fix the ozone layer, as was the case with the Montreal Protocol, but this is where we are and it is thus time to act.

A brief history of climate negotiations

Since 1988 there has been a series of key events in the United Nations (UN) and national efforts against climate change leading up to the most relevant deal of our times: the Paris Agreement. Over 190 countries, scientists, non-governmental organisations (NGOs), change makers and some involved citizens since then have been building these stepping stones and final agreements on behalf of us *all* . . . blah-blah-blah – you're probably thinking: *Why are all these agreements relevant for my own individual change, and why so much about the UN?* The root of how we, as a society, need to change is embodied in a huge organisation such as the UN orchestrating not only national governments but also NGOs, all the way down to civil society conforming to goals and targets that connect all of us for a better standard of living and a healthier planet. This means that if we know about these activations, we, as individuals, can be part of these movements.

The Earth Summit in Rio de Janeiro in 1992 was the first summit that really aimed to support socio-economic development and prevent the deterioration of the environment, and to lay a foundation for a global partnership between industrial and non-industrial countries. Since then a series of annual negotiations have proceeded.

1992, Rio de Janeiro, Brazil World leaders gathered for the first Earth Summit and signed the United Nations Convention on Climate Change – the first international treaty aimed at limiting greenhouse gas emissions; however, these commitments were not legally binding, resulting in nothing more than a friendly chat between countries.

Since 1995, Conference of Parties (COP) was established and the first COP1 took place in Berlin, Germany. The conference voiced concerns about the adequacy of countries' abilities to meet commitments, and highlighted how each member of the UN Convention on Climate Change had to shoulder their own responsibilities to reach the decisions needed to promote its effective implementation. Since this, joint measures and COP summits have occurred globally on an annual basis to review the communications and emission inventories by all countries that are part of the convention against climate change.

1997, Kyoto, Japan The Kyoto Protocol was adopted, setting binding emissions targets for wealthy countries. The United States didn't join this, though, because it said it would further harm the US economy as it was facing a downturn at the time, and an energy shortage, and it could not imagine a foreseeable future without fossil fuels. The other reason: the US accused the protocol of not being fair, because developing nations such as China, India and Mexico were given no targets and allowed at their free will to increase emissions if desired.[13] Although the Kyoto Protocol received significant pushback from global superpowers, it did set a precedent for future climate negotiations to be fairer on *all* countries reducing their emissions.

2009, Copenhagen, Denmark This was the first attempt to craft a global emissions treaty to replace Kyoto, which was set to expire in 2012. Negotiations fell apart amid disputes between rich and poor countries over who should do what, but it ended with a voluntary deal inviting countries to present non-binding emissions targets for 2020. Quite frankly, Copenhagen was a failure.

2011, Durban, South Africa UN climate talks produced a major breakthrough, as countries agreed to adopt a universal agreement on climate change in 2015 that would take effect five years later and apply to all of them.[14]

Note: it is important to mention that a complicated ramification of many of the UN agreements is that they result in gaps between the signing of documents and the actual ratification and implementation. The time lag can often take years from the conference until countries start to act.

2015/16, Paris, France More than 190 governments met to finalise an accord that would keep the average global temperature increase below 2°C and curb greenhouse gases once and for all.

2015 – *the* year

The most relevant year in the climate space to date is 2015, because of the Paris Agreement and the introduction of the UN Sustainable Development Agenda.

The Paris Agreement

A landmark environmental accord that was adopted by nearly 200 countries in 2015, the Paris Agreement addresses climate change and its negative impacts. The deal aims to substantially reduce global greenhouse gas emissions in an effort to limit the global temperature increase in this century to 2°C above pre-industrial levels while pursuing means to limit the increase to 1.5°C.

The agreement includes voluntary commitments from all major emitting countries to cut their climate-altering pollution and to strengthen those commitments over time. The pact provides a pathway for developed nations to assist developing nations in their climate mitigation and adaptation efforts, and it creates a framework for the transparent monitoring, reporting and ratcheting up of countries' individual and collective climate goals.[15] In contrast to other agreements that might be low on targets and selective participation, the Paris Agreement is detailed, pragmatic and fairer for all parties involved and is what can now be labelled the definitive universal climate accord. It aims to:

- Limit global temperature rise by 2°C by reducing greenhouse gas emissions. Beyond 2°C we risk dramatically higher sea levels, changes in weather patterns, food and water crises, and an overall more hostile world.
- Provide a framework for transparency, accountability, and the achievement of more ambitious targets. The agreement is currently voluntary, so it doesn't tell countries exactly how

they should do things and there is no defined punishment for breaking it.

- Encourage richer countries to help out poorer countries by mobilising support of up to $100 billion per year for climate change mitigation and adaptation in developing nations.

The situation today

How are the participating countries, including the UK, actually doing in sticking to the agreement five years later? Unfortunately, many large emitters are not on track to meet their self-imposed targets. What's more, even if every country did manage to fulfil its individual pledge, the world would still be on stride to heat up well in excess of 2°C. The last four years were the four hottest years on record, and we used to think that if we could keep global warming below 2°C this century, the changes we would experience would be manageable. Not anymore. Later studies say that going past even 1.5°C is dicing with the planet's liveability[16] and the world is completely off-track, heading towards 3°C. To reach the broader Paris goals, countries must dramatically accelerate the transition towards clean energy over the next 12 years, but each year emissions are on pace to rise sharper, so, in plain English – we are way off course.

Panic? Well yes, but not entirely. The harsh reality is that countless pledges are no longer sufficient to save our planet. Citizens need to see actual action and the implementation of these promises, and only when we know what these promises are, can we hold our governments accountable for climate action. If the people who agreed the Paris negotiations can't act upon on their own goals, by retiring coal plants in favour of renewables, or improving transportation, it requires us as individuals to start pressuring governments by doing our own thing to help the agreement succeed. We should be getting angry with the people with more power who are not using their voice for inspiring change. That could be as small as considering our own energy consumption and switching to lower carbon alternatives.

The Sustainable Development Agenda

In 2015, all countries of the UN adopted the 2030 Agenda for Sustainable Development. At its heart are the 17 Sustainable Development Goals (SDGs), which are an urgent call to action by all countries, both developed and developing, in a global partnership for peace for people and the planet. The goals recognise that ending poverty and deprivation must go hand in hand with strategies that improve health and education, reduce inequality, and spur economic growth – all while tackling climate change and working to preserve our oceans and forests. It is a commitment to eradicate poverty and achieve sustainable development by 2030 worldwide, ensuring no one is left behind. The adoption of the agenda was a landmark achievement, providing for a shared global vision towards development for all, broken down by the 17 Sustainable Development Goals (SDGs) and 169 targets.

The 17 Sustainable Development Goals (SDGs) to transform our world:

Goal 1: no poverty
Goal 2: zero hunger
Goal 3: good health and well-being
Goal 4: quality education
Goal 5: gender equality
Goal 6: clean water and sanitation
Goal 7: affordable and clean energy
Goal 8: decent work and economic growth
Goal 9: industry, innovation and infrastructure
Goal 10: reduced inequality
Goal 11: sustainable cities and communities*
Goal 12: responsible consumption and production*
Goal 13: climate action*
Goal 14: life below water*
Goal 15: life on land*
Goal 16: peace, justice and strong institutions
Goal 17: partnerships for the goals

* Goals 11 to 15 are directly related to the climate.

The 17 goals rely on a healthy planet

There is rarely unilateral agreement among most countries in the world over a single topic, but with the Paris Agreement and the Sustainable Development Agenda, heads of state from around the globe collectively agreed that climate change driven by human activity is a threat so great to humanity that all must be part of the solution. Agreements such as these take years to negotiate. In order to make the 2030 Agenda a reality, broad ownership of the SDGs must translate into strong commitments by all stakeholders to implement the goals.[17] For the first time in history, the topic of climate change was high on the agenda for those writing the goals and, as a result, over one-third of the 17 goals have a climate focus. To take it one step further it would not be bold to say that without a healthy planet none of the other goals will succeed.

Making the goals understandable for all

One of the biggest frustrations with previous UN protocols has been the legal, technical and formal jargon used in these agreements that is not easily translated to the everyday person. How can nations and global institutions say they are working for 'all' if they are not including everyday people and youth in their negotiations or implementation phases? The Sustainable Development Agenda has done a phenomenal job in trying to change that stigma through the UN and countries hosting events in honour of the goals, and even businesses – large corporations, too – are using the agenda and goals to structure their businesses moving forward. There are also continuous conferences and discussions, social media campaigns, and celebrity involvement in trying to boost awareness about the agenda and its goals. As a result, every day more nations, corporates and individuals are making new commitments, pledging new financial structures on our behalf around climate change and sustainable development.

Nevertheless, not enough people know what the Paris Agreement entails or what the 17 Sustainable Development Goals are. The UN is one of the most powerful allies we have to fight climate change, and if we have this sustainable framework we must amplify it to our family and friends as much as possible. The sustainable agenda basically says that we must meet the needs of the present without compromising the ability of future generations to meet their own needs, and if it encompasses society, environment, culture and economy, the future is bright. The only way we can do that is by becoming aware of our consumption patterns and making changes. We could go on to discuss the potential scenarios facing more people and the planet but that would be contradictory to the core purpose of this book. Through a combination of urgency and change, we, the people, have the opportunity and a responsibility to play a role in saving the planet and to make a difference and allow us to keep that optimism. We have finally reached an era where we have clear guidance and roadmaps of what to do and how to choose the right path. Everything from the future of business with purpose, sustainable principles incorporated into government policies, and our own lives awaits us. Do not allow a lack of knowledge or a sense of hopelessness to cloud your potential. We are vaguely aware of when we choose to not look at the facts and evade the unknown, but we must embark on a moral revolution.

HOW CLIMATE CHANGE CAN DEVASTATE A COMMUNITY

Recognising the interconnectedness of the sustainable development goals is imperative, and here is a story of why:

One of my first ever climate field projects in 2017 was to a small village called Anse du Clerc on the most westerly tip of Haiti in the Caribbean. With a population of fewer than 10,000 villagers, this community is isolated off the beaten track, hours walk away from any local market or any sort of relief. Most people live in makeshift concrete-and-tent shanty structures. The community for decades had thrived and sufficed off the fishing industry both for food and income.

This dramatically changed after the devastating 2010 earthquake that destroyed huge areas of Haiti, including this village, but rising sea temperatures and levels, plastic pollution and illegal over-fishing are leaving the community in even more despair. More powerful storms are hitting their community more frequently, affecting clean water supplies, and they are engaged in a constant battle with their only asset: the ocean.

The carbon footprint of Anse du Clerc is small, as most people do not even have access to a regular electricity supply, and this is just one example of how a community most vulnerable to climate change is among those who contribute towards it the least.

The people of Anse du Clerc now fish only to survive. The

fishermen can no longer fish in the shallow waters around the island because of a dramatic decrease in catch. Now, the fishermen must travel for up to 10 hours to the deep sea on small and rickety wooden boats, which is dangerous because of unpredictable weather patterns (Haiti's coastline makes it particularly vulnerable to hurricanes). When they return, they often have enough fish only to feed their own. Without enough fish to also sell at market, there is no income and an overall increase in poverty levels within the community. Without income the health and well-being of community members becomes vulnerable; and children are often not sent to school because their parents simply cannot afford to send them. If some are chosen to go, it's normally the boys rather than the girls, which increases the gender equality gap and reduces access to quality education.

If the environment prevails, then life can go on for vulnerable communities globally. There is time to mitigate and adapt using the sustainable development goals as a framework, but climate change is an issue of the future that we have to think about and plan for now.

It is important to contextualise Haiti: from a historical perspective the island was rooted in slavery and following a brutal war of liberation, its historical context has made it an incredibly poor country. Not even today does a city in Haiti have a regular electricity supply – for many residents, wood remains the most important source of energy and one of the reasons why the island's forests have largely disappeared. Without vegetation, soil is not kept intact, and the exposed ground makes it easy for heavy rains to wash away the soil cover, which makes it harder to grow crops.[18] When climate change hits Haiti, it hits the island hard, and although the example of Anse du Clerc might be far away from you, it is just one story of many globally. There are many communities closer to home whose future depends on the stability of our climate. Open your eyes and look around.

'Climate change is the single biggest threat to vulnerable people in the Caribbean, causing over $100 billion in damages over the past 20 years alone. The impact on individuals is even more devastating when people lose everything in just a few seconds during intense storms. These frequent unnatural weather patterns in the Caribbean are a reminder of how climate change multiplies the power of natural disasters. These disasters disproportionately affect the poor living in areas of weak infrastructure, who have limited ability to cope with the impacts. Never before has something like the Sustainable Development Agenda offered promise of so much for humanity, and we must strive to implement all 17 goals by 2030 to ensure future generations a fair and prosperous future.'

Laurent Lamothe, former Prime Minister of
Haiti and visionary social entrepreneur

2

Plastic Service Announcement

'Only we humans make waste that nature can't digest'
Charles Moore, marine researcher

Plastic is cheap and convenient, and, thanks to this material, countless lives have been saved in the health sector, safe food storage has been revolutionised and the growth of clean energy, from wind turbines to solar panels, has been greatly facilitated. This chapter is not by any means suggesting that we humans must go without plastics, but I'm simply saying that we have to do things differently.

A brief history of the plastic bag

It has been 54 years since the plastic bag was created, and it is now one of the most commonly found items in our oceans and washed up on beaches globally. Here is a brief history:

1933, polyethylene The most commonly used plastic was created by accident at a chemical plant in Northwich, England. The chemical composition had been created in small batches before and was initially used by the British military during World War II.

1965, Swedish engineer Sten Gustaf Thulin designed for the company Celloplast the one-piece polyethylene shopping bag, and it quickly began to replace cloth in Europe. The company obtained a US patent for the idea that was later called 'the T-shirt plastic bag' and its design, essentially, is used for every plastic bag you have ever been given in a supermarket.

1979 Controlling 80 per cent of the bag market in Europe, the plastic bag was sent abroad and quickly became popular in the United States. Plastic companies began to aggressively market their product as being far better than paper and reusable bags.

1982 The largest supermarket chains in the United States switched to plastic bags, and by the end of the decade plastic bags had almost replaced paper around the world.

1990s Plastic bags were now a universal product manufactured at a rate of one trillion a year. They were showing up around the polar ice caps, in the darkest depths of our oceans and clogging waterways for communities globally. Numerous wildlife species mistake plastic bags floating in the ocean for food.

2002 Bangladesh is the first country in the world to implement a ban on plastic bags, after it was found that they played a key role in clogging sewage systems during disastrous flooding. Other countries began following suit. Africa took a serious stance with more than 15 countries, such as Kenya and Uganda, completely banning plastic bags or heavily taxing them.

2015 The number of single-use plastic bags peaked at 7.6 billion. In the UK in 2014, the government, supported by environmental groups, enforced a 5p tax on bags. Within just one year, studies found that 1.1 billion single-use plastic bags were sold – a massive drop over the previous annual total.[1] That figure didn't include the stronger plastic 'bags for life', which were being purchased instead of the single-use bags because supermarkets are mostly only required to give figures for the single-use plastic (although some do give both). In 2019 shops sold almost 958 million 'bags for life' in just one year and the average UK household used 44 bags for life in that one year. Each of these 'bags for life' can weigh as much as four times a single-use bag, grossly thicker and less biodegradable. If we balance out the single-use with 'bags for life', the evidence for a dramatic decrease in the sale of bags in general is not as compelling an argument as one may assume.

2020 About one trillion single-use plastic bags are used annually across the globe. That is nearly 2 million every minute, a number that is still far too high. Modern shopping bags are increasingly recyclable, or biodegradable made from vegetable-based plastics, which can decay organically and prevent a build-up of toxic chemicals in landfills of plastic. If you are not bringing your own reusable bag to the shops, opt for biodegradable or paper instead.

Plastics and the climate discussion

As much as we might try to avoid it, plastic for most of us is part of everyday life.

The controversy over plastic is everywhere, but one thing needs to be said: plastic is not the be-all and end-all of the environmental movement and it rarely actually comes up in climate discussions. Organisations have categorised plastic pollution as a thing of its own, but plastic is at the core of climate change, because plastic is a by-product of fossil fuel: those greenhouse gases that are warming our planet and fucking shit up.

Since the 1950s, the production of plastic has outpaced that of almost every other material. Much of the plastic we produce is designed as single use, mostly for packaging, and it is intended to be thrown away after being used only once. There are many types of plastic materials – such as polyethylene, PVC or acrylic, to name some of the most popular – and they have incredibly useful and durable properties.

We are a culture that treats plastic as a disposable material rather than a valuable resource to be harnessed. We are also at a point in humanity where at the supermarket you can buy one single banana on a bed of polystyrene wrapped in plastic. Yes, wrapping fruits and vegetables in single-use plastic packaging is totally bananas. In fact, so many items come unnecessarily wrapped in plastics, and for those reasons the plastic statistics go so far off the scale that it really is hard to put the numbers into context. Our oceans and waterways are choking. Plastic soup patches are cropping up globally. Marine life and sea birds are dying at an unprecedented rate. In cities, plastic is clogging drains, causing floods and breeding disease. Plastic pieces are found blocking the airways and stomachs of hundreds of species, often ingested by birds, turtles and dolphins who mistake plastic for food, and it ultimately ends up in our own food system. We are harming our own human health through this epidemic.

The hard facts: plastic

- Three-hundred and thirty million tonnes of plastic are produced every year.
- Over 12 tonnes of plastic enter our oceans every year.
- One rubbish truck of plastics enters the ocean every minute.
- The US and the UK toss out over 550 million single-use straws every day.
- Eighty per cent of plastic in the ocean originated on land.
- Plastic could outweigh fish in the ocean by 2050.
- It takes 450-plus years for one plastic bottle to break down in the ocean.
- Ninety per cent of seabirds are likely to have plastic in their stomachs.
- Around 500 billion plastic bottles are sold every year and growing – that is the equivalent of 20,000 per *second*.[2]

- The US alone uses 1,500 of those plastic bottles per second.
- Japan is the second in the world (after the USA) in terms of plastic packaging waste per capita. According to Statista, Japan produces more plastic per capita than China and the rest of Asia combined.[3]
- Coca-Cola produced three million tonnes of plastic packaging in 2017, which is the equivalent to roughly 200,000 bottles a minute.[4]

THE WORLD'S MOST POLLUTED RIVER

Words will never be enough to describe and contextualise what some devastated places in the world look like. The destruction is just unfathomable to the naked eye, and one of those places is Indonesia's Citarum River, aka the world's most polluted river. Every day, no less than 20,000 tonnes of waste and 340,000 tonnes of waste water, mostly from 2,000 textile factories, are discharged directly into the once clear and pristine waterways of the Citarum River. No wonder the fish are mostly gone from the river.

To visualise how dirty the river is, at some points the water surface is invisible to the human eye, because it is obscured by an unbelievable amount of rubbish, mainly plastics. When there is a glimpse of the water, the colour is murky and dark due to the excessive volume of toxic chemicals being dumped into the river by neighbouring industries. Other sources of pollution are the smaller villages lining the Citarum River, which have no public rubbish collection or landfills, leaving residents to choose between burning their own waste and sewage or throwing it in the river. Lastly, lots of the plastic pieces are also waste from Europe and the USA that has been transported to Indonesia for recycling, as I explain on page 61.

The Citarum River is a lifeline for many people in the country. It is the source of 'clean water' to over 25 million people for drinking and agriculture across Indonesia. It will be no surprise that the Citarum has never met the water quality standards of the Indonesian government. The many people that depend on this heavily polluted water now suffer from infections and skin diseases.[5]

This is not a problem unique to the banks of the Citarum, and the plastic polluting the waterway doesn't just end there: it floats down into the ocean. If we don't stop polluting all our water sources, it is estimated that there will be more plastic than fish in our oceans by 2050.

Plastics 101

Have you ever taken a moment to think about where plastic – a material you grasp in your hands, drink from, eat from and use for many other intimate situations – actually comes from?

As mentioned earlier in this chapter, plastic is made from fossil fuels. Yes, fossil fuels. What is one of the biggest challenges in helping save the planet from plastics? The petrochemical companies, of course.

I'm going to get geeky here, but trying to simplify the plastic production process is complicated, so this mini explanation will be missing out hundreds of details – but take what you can for the 101 basics. (And for those really interested in the chemistry of plastics, there are literally thousands of hours of online material about plastic production on Google – knock yourself out.)

Plastic is derived from oil. Crude oil is a complex mixture of thousands of chemical compounds and is heavily processed before it is used. The life of plastic starts with the distillation of crude oil in an oil refinery, which groups the oil into different fractions. Each fraction is a mixture of hydrocarbons (carbon and hydrogen), which differ in size and structure. The fraction necessary to produce plastic is called naphtha.[6] Polymerisation and polycondensation through specific catalysts are the two main processes used to produce plastics. What makes plastic so versatile is that it is not just one material but a group of materials – and these make up so much of our lives.

Big business before pollution

In discussions, public-awareness campaigns and high-level political forums, rarely do we see or hear about the petrochemical companies. Plastics manufacturers assume demand for disposable plastics will continue to rise, despite the evidence that global awareness of plastic pollution is growing, and cultural attitudes are changing. Industry investments reflect a further underlying assumption that supplies of cheap hydrocarbons will remain the norm for decades to come. Companies are pumping billions of dollars into new state-of-the-art facilities producing the raw materials for plastic making. These powerful billion-dollar corporations have every vested interest in keeping plastic production alive without giving a single crap about people or planet. This makes the fight against single-use plastic pollution more compelling and holistic, realising that good choices in renewable energy and climate-friendly decisions might also help to reduce single-use plastic production and pollution, and vice versa.[7]

The consequences

The miracle child of the 1930s was durable, flexible and cheap, and it quickly became part of everything. At the time, as with thousands of new inventions, the long-term consequences were never evaluated and, as a result, we face a plastic crisis – from waterways to the hidden depths of our food chain. Even everyday foods contain plastic. Here are some examples:

Plastic bottles

Most bottled water is sold in disposable plastic bottles that leach microplastics and other chemicals into the water contained therein. Bottled water is often sitting in plastic for months, having been filled just after the plastic bottle was created, and it is off-gassing chemicals. The best water to drink is filtered tap water, or try buying glass-bottled water instead.

Tea

Around the world, 2 billion people will drink tea every morning. As a Brit, tea is a staple daily drink, and as a nation we drink on average 165 million cups of tea every day.[8] An estimated 96 per cent of the tea bags we use include plastic: polypropylene. Companies started using plastic to seal the tea bags, making it completely polluting every time we dispose of one in our bin, and bad for our health of course. Next tea break, look out for plastic-free tea bags or choose loose tea.

Coffee

Over 30 billion plastic and aluminium coffee capsules have been sold by Nespresso alone to date. The world has placed an urgent need for coffee above the needs of our planet. If you use conventional coffee capsules, consider opting for compostable plastic-free capsules, or simply switch to French press, filter, traditional espresso (no waste apart from grounds) or instant that does not require plastic to be discarded each time it's used.

Tinned food

Most cans are lined with chemical BPA that contains traces of plastic that has been linked to cancer and obesity; be wary because the plastic lining is often a transparent lining. The plastic leaches into the contents of the can and is also a source of microplastics. If available and possible, buy glass jars instead, or if you are able to shop in a waste-free store you can fill your own full of pulses and other good stuff.

Sea salt

As the seas are gradually getting more polluted, it is no surprise that sea salt is often contaminated. Studies have shown that 1kg of sea salt can contain over 600 microplastics.[9] Choose high-quality rock salt that was formed before the world was polluted.

Beer

There have been several studies on beer that found microplastics in the final products. The source of these microplastics is attributed to the water supply, as they used municipal water. Read the beer label and only drink beer that was made with filtered water.

> 'Plastic will be the main ingredient of all our grandchildren's recipes'
>
> Anthony T. Hincks

Action against plastic

We are never going to solve the plastic crisis overnight, as the sheer amount of plastic that exists and is currently being produced is overwhelming. Our dependency on the material is simply too great. Like climate change, the plastic pollution debate is overwhelming and huge. The only way we are going to make a start on solving this problem is to stop producing and using single-waste items at the scale that we are and to find alternatives to wrapping and our culture of convenience. In order to change we sometimes need a little inspiration, and community leaders tackling plastic pollution are on the rise, whether it be coordinating a beach clean-up or organising a community gathering to talk about plastic issues.

HOW PENZANCE BECAME BRITAIN'S FIRST EVER PLASTIC-FREE TOWN

Plastic pollution has become synonymous with most coastal activity across the UK: on the beach, sailing, wildlife spotting ... spend time on or near the sea and it's obvious that our coastlines are plagued with plastic – they're just part of our experience. One town that decided to take action to minimise its own contribution to the plastic crisis is Penzance, in Cornwall. It became the first town across the country to receive 'plastic-free' status from the marine conservation charity Surfers Against Sewage (SAS), after the town hall voted to support the initiative. Shops, cafes and other local businesses have reduced single-use plastics by switching to biodegradable alternatives, and community members got involved to clean up the beaches.

Having formed a plastic-free community steering committee, leaders of the town are committed to taking the battle up the chain so that consumer pressure compels more businesses and even the suppliers to change practices, reducing the cost of being plastic-free. Not that scepticism is absent in Penzance, but conversations in the town demonstrate change is not a luxury, with both businesses and residents' perceptions over the issue turning the tide. Since Penzance achieved plastic-free status, more than 330 communities across Britain have applied to follow the town's lead.

One plastics activist in Kenya has risen to fame and become a national hero. Fed up with what he was experiencing in his town, and concerned for the fate of his son's generation, James Wakibia from Kenya decided to act. He shares his story:

James Wakibia
Plastics activist, Kenya, @jameswakibia

Q How was plastic affecting your hometown and community?
A Plastic bags were strewn almost everywhere, every space and

corner had plastic bags; they blocked drains, clogged sewer lines, they were trapped on trees and hedges, they were a menace. They also held water when it rained and that became home to mosquitoes, which causes malaria.

Q What was the turning point and moment you decided that you had to act and become an environmental advocate?

A My initial calls to demand the closure and relocation of my town's landfill, which was a complete eyesore, unfortunately failed. I did have promises from the county government that they would act in the near future, but while waiting it dawned on me that the biggest problem with the landfill was not general waste but plastic, which is never recycled, which never biodegrades, is flimsy and therefore can easily be transported by wind and water, unlike other wastes, which can biodegrade and can be recycled. From then on, I decided to run campaigns calling for bans of plastic bags. That was from 2013 until 2017, when the government of Kenya finally banned plastic bags.

Q What tactics did you use to raise awareness about your plastic campaign? How important was social media to amplify your message?

A I tried practically every available means at my disposal to pass my message to the general public. From the word go, I wrote opinion articles and published in Kenya's mainstream media, I also published on blogs and social media. Later I had a great liking for Twitter and Facebook: Twitter more so because I was able to track the hashtags and see how many people I was reaching. Also, going viral on Twitter was such an amazing thing because of the numbers and the discussions that followed. Indeed, it is on social media that a government official, the Minister for the Environment, responded to my then trending Twitter hashtag #BanPlasticsKe before changing the hashtag to #IsupportBanPlasticsKE. After the Environmental Minister said that she supported calls to ban plastic bags in Kenya a new law was quickly implemented. Social media is such a powerful tool for change, because it has no boundaries, it's instant, and it can reach anybody anytime. Social media has been used to remove tyrants from power – it's a medium of communication that is like no other. Greta Thunberg is using it to fight the climate crisis, I used it to fight plastic pollution in Kenya and I will always use it for positive change.

Q What are your ultimate goals for both your hometown and Kenya as a whole in its crusade against plastic?

A My ultimate goal is to see that we as humans live in a sustainable world where everything we use can easily be recycled, or up-cycled, and which have very minimal effects on the environment. If a thing that we have manufactured cannot be recycled and at the same time cannot biodegrade, we should stop using it; it is not rocket science that that thing will haunt us or our children some day. I wouldn't want that for my son, who is two years old. I want him to live and enjoy the environment without worrying whether the water he drinks is contaminated with plastic or the fish he eats contains microplastics. That's my dream for him and all the other kids like him.

I will be happier to see more countries doing away with single-use plastics for better alternatives. There is no life in a world choking from plastic waste; it is time we removed the blinkers, now more than ever I feel the world needs to change.

This planet is the only home we have, we must jealously protect it – less plastic is fantastic.

Recycling

The main problem is that plastic never disappears – it breaks down. We've been told to *recycle, recycle, recycle* by governments and media alike, but let's debunk a myth. In the common phrase 'refuse, reuse, recycle', the third R can be incredibly deceptive, as recycling process systems are not that far advanced. Our ability to cope with plastic is already overwhelmed, and less than 9 per cent of the 9 billion tonnes of plastic the world has ever produced has been recycled.[10] Recycling practices vary from country to country, but even in the most advanced nations, such as Japan and Denmark, the plastic pollution crisis has far exceeded any country's ability to cope. In the past 20 years alone, it is estimated that over 350 million tonnes have been produced, but what is overwhelming is that figure is set to double across the next two decades. From the first ever item created to where we are now, the upward production curve is essentially off the charts.

Most of the items you put into the recycling bin, due to certain imperfections, are unable to be processed. When different materials are mixed together, the weakening of the material and a variety of other technical issues make the reuse of plastics challenging and, thus, not much progress has been made in effective recycling.[11] When too many of these types of items end up in the recycling bin the contents are labelled 'contaminated' and are unable to be recycled.

And so, if most plastic is not being recycled, where do you think all the plastic goes? It doesn't just disappear off the face of the Earth or evaporate into thin air. It must go somewhere, and that somewhere is either in a landfill, into our oceans, or to a poorer country.

Landfill

If present trends continue, by 2050 there will 12 billion tonnes of plastic in landfills. That amount is 35,000 times as heavy as the Empire State building.[12]

A landfill site is an area for the disposal of waste materials by burial. To put it simply, landfills operate by layering waste into a large hole. Historically, landfills have been the most common method of organised waste disposal and their purpose has been to isolate rubbish from its surrounding environment, preventing water contamination and contact with air. Plastics mixed in with general waste are part of growing mounds globally producing horrendous odours from gases such as sulphides, methane, carbon dioxide and ammonia, among others.

Landfills are not designed to break down waste, only to store it, but rubbish is decomposing, albeit slowly in a sealed way, and the tonnes of methane and carbon emissions released into the air contribute to greater global warming.

Incineration

From an environmental perspective, it is generally deemed better to bury plastics than to burn them, but six times more waste is burned in the US than is recycled, for example. The most dangerous emissions can be caused by burning plastics containing toxic chemicals that are extremely harmful to human organisms.

Because plastic is made from oil and gas, it creates a lot of heat when it is burned; the hydrocarbons inside the material are actually more energy-dense than coal. Converting plastic waste into energy does nothing to reduce demand for new plastic products, or significantly reduce CO2, but it could be part of the circular economy if the process could become cost-competitive globally.

Our oceans

Working out exactly how much plastic is in the sea is challenging, considering humans have explored less than 4 per cent of the oceans. Also, the ocean is constantly moving, so plastic travels far and wide. A study by the Ellen MacArthur Foundation estimated that 150 million tonnes of plastic are currently in the ocean. An estimated 80–90 per cent of plastic in the oceans comes from land-based sources. Half of all that plastic comes from fishing nets and gear that gets lost from trawlers. Plastic fishing nets are the most dangerous for sea life such as dolphins and sea lions, which get trapped in the nets and can easily suffocate.

PLASTIC DUCKS TOUR THE WORLD

In 1992, 29,000 yellow rubber ducks were plunged into the ocean after a shipping crate was lost at sea on its way to the US from Hong Kong. After being lost in the Pacific Ocean, the ducks made it around the world, washing up on the shores of Hawaii, Alaska, South America, Australia and Europe. Others travelled over 17,000 miles, floating over the site where the Titanic sank, and spending years frozen in the Arctic ice, to eventually reach the US.

The disposal problem

The idea of cleaning up so much plastic is mind boggling; furthermore, what would humans do with so much plastic if we dragged it all out of the ocean? We have nowhere to put it, not enough ways currently to recycle it, and unjustly it most probably would end up burning on a landfill in a developing country. Clean-ups can have limited success. In coastal areas where plastic is so bad, it can threaten the tourism industry and infrastructure in communities. Beach and river clean-ups are vital, even though there are *disposal* problems with the plastics, because clean-ups contain the plastics in one place – and they are a great way for you to get involved in your local community.

What happens to the plastic that doesn't get taken out of the ocean? Instead of disappearing, it slowly breaks down into smaller fragments known as microplastics. The main issue with microplastics is that they filter into our water systems and food systems. Hundreds of thousands of marine creatures, and especially sea birds, die from plastic entanglement or ingestion. As I said at the beginning of the book, if you eat fish, there is a 99 per cent chance you have already ingested microplastics.

WHAT ARE MICROPLASTICS?

Microplastics are tiny pieces of plastic, sometimes defined as ranging from 5mm to 10nm (nanometres) long. Some microplastics, such as microbeads, are designed to be that small (see page 68); however, most microplastics are fragments of a larger body of plastic, such as plastic bags or bottles, that are breaking down into ever-smaller pieces.[13] The main source of microplastics, though, are said to be released through washing our clothes, and those account for roughly one-third of the plastic in the ocean. We will touch more upon the issue of microplastics in the fashion chapter.

Plastic waste and poor nations

For too long, both as entire developed countries but also as individuals, we have avoided taking responsibility for our own plastic consumption. Although we, in the developed world, might be the largest polluters of plastics, these stats don't make it to official figures because countries have never had to monitor officially where waste goes once it leaves their borders.

For years, developed nations have been shipping hard-to-recycle plastic waste to poorer countries across Asia and Africa. These plastics are coming from an array of industries, such as food and beverage, healthcare, fashion, technology and aerospace. Instead of finding ways to deal with its own waste across Europe and America by innovating throughout recycling systems, it was decided that the easier and 'better' solution was just to ship the unwanted plastics to other countries, which have fewer resources and won't complain, and let them deal with it. The idea was that poorer nations would be able to repurpose international plastics – but the reality is far from successful. The less valuable plastics that are harder to recycle almost always end up discarded rather than even being attempted to be turned into new products by anyone in any country. As a result, plastics from America and Europe are found piled up in villages throughout developing nations. The international plastic debris that clutters foreign lands, floating in the oceans and endangering wildlife, is creating dire consequences for the people and communities on the receiving end of our waste shipments.

One of the main recipients of discarded plastic was always China, but towards the end of 2017 it, too, had finally had enough and announced that it would no longer be able to accept waste from the West because its systems were not able to cope with its own growing domestic plastic production. China also saw an opportunity, given the current political climate, to stand up and meet some of its own climate targets. This became catastrophic for Europe and America, because they were so used to shipping away their plastic problems.

This left governments in a major pickle, but it was also an opportunity for the West to create real, system-wide change around plastic production and consumption. However, instead of adopting alternative strategies and improving plastic waste strategies, governments increased exports to other developing countries such as Indonesia, Vietnam and the Philippines. These countries quickly became overwhelmed with shipments of waste that they do not have the capacity to handle. Helping poorer countries deal with the ever-increasing amounts of plastic and other waste is ultimately a core development issue.

Myth buster: recycling is only for plastics

Recycling is the separation of specific materials, commonly including plastic, metal, paper, juice and milk cartons, cardboard, glass, and even empty aerosols, and diverting them to a facility that will repurpose them for reuse. To guarantee that all the items tossed in your recycling bag or bin are recycled, rinse them out, wash and clear all plastics of food residue, because dirty plastic can't be recycled.

Recycling at home completely depends on where you live and what your local council accepts. Check in with your council for a full list – you should be able to find what you can and can't recycle on your council's recycling page online. For now, here is a short reminder of non-conventional items that may or may not be allowed in your recycling bag:

The do's:

- Junk mail.
- Empty aerosols, aluminium, steel (tin) cans.
- Clean shampoo bottles, detergent bottles and plastic cleaning containers from the kitchen, bathroom and laundry.
- Only green, brown or clear glass bottles.
- Clean aluminium trays and foil.
- Empty yogurt, margarine and plastic takeaway food containers.
- Paper waste such as magazines and copy paper.
- Cardboard – yes to juice cartons, tissue boxes, milk cartons. Only *clean* cartons can go in recycling.
- Lids? Big ones from ice cream tubs are big enough and metal ones from jars. The small ones from plastic bottles normally don't make it through the process.

The don'ts

- Don't put plastic bags and other soft plastics in the recycling, as they create problems at the recycling centres. They can wrap around the machinery, causing shutdowns.
- Disposable coffee cups can't go in the recycling bin. Many coffee cups are wax- or plastic-coated and can't be recycled, so double check with your favourite coffee shop – or, better still, take your own reusable cup.
- Lids from hard plastic bottles, beer bottles, and so on.
- Black and brown plastic containers.
- No polystyrene foam packaging – no takeaway foam containers or those polystyrene bits that things came floating in.
- No clothing or rags of any kind.
- No tissue paper or kitchen roll when it has come into contact with food, grease or bodily fluids.
- No nappies.
- No greasy, dirty pizza boxes.
- No light bulbs.
- No shredded paper.
- No garden or food waste.
- No homeware – even though glass can go in the recycling bin, drinking glasses, oven glass and mirrors are not recyclable.
- No wrapping paper that is foil, glittery or somewhat shiny. This is not recyclable and cannot be processed because recycled paper would end up with glitter in it.

Where are we now?

Demand for plastic is set to double over the next 20 years, and if habits and production modes don't change, there could be more plastic than fish in the ocean by 2050. Thanks to wide-scale reporting of the consequences of the widespread use of plastic, high levels of activism, social-media pressures and environmental groups, the plastic crisis is definitely on the global agenda. A recent declaration made by 193 countries in December 2017 in Nairobi, Kenya, highlighted the urgent need to tackle the plastic crisis. This was, of course, another UN declaration, but, as we learned from the introduction, getting that many countries to agree on a single topic and the need for action is nothing short of an achievement. It marks the first global pledge to tackle the plastic crisis head on.

In some places, banning single-use plastics is a powerful message and creates a sense of urgency in communities to change behaviour rapidly and incentivise local companies to make alternatives. Over 60 countries have issued bans and levies on single-use plastic, which is a great start, but implementation is where the real work begins. All governments should be introducing financial incentives to change the habits of consumers, retailers and manufacturers, enacting strong policies that push for a more circular model of design and production of plastics. In a circular economy, the final destination of plastic would not be landfill or incineration. Instead, the plastics would be designed to be reused at their highest economic potential for as long as possible within the global supply chain.

> 'Plastic isn't the problem. It's what we do with it. And that means the onus is on us to be far smarter about how we use this miracle material'
>
> **Erik Solheim, Head of UN Environment**

Leaders in reducing plastic pollution

There are good examples of corporations or companies that have played a role in tackling plastic pollution. The likes of Adidas, Lush and Patagonia have demonstrated that the private sector will not stand idle, and have launched new products made from recycled plastics. In addition to great marketing campaigns, companies have reaped economic benefits by transforming certain business practices for good along global-value chains, and they have found that linking to sustainable consumption and production has been hugely profitable for them. (There is, however, a flip side to making new clothing from recycled plastics, as I explain on page 88.)

Plastics are fundamental to our everyday life, and catalysing change through partnerships in the global materials sectors will not only create more effective waste systems but also demonstrate the potential shift away from a linear to a circular economy. The mass production of single-use plastics is a recent issue, which is now, thankfully, at the forefront of political agendas. If companies are not ready to phase out plastics on scale immediately, they do need to be considering this from the very beginning – and during the second, third and multiple lifetimes of the products we use. We then need the efforts of creators and companies to redesign packaging and to innovate using alternative materials.

There are various stages, but a common vision for the war on plastics is best summarised by a study conducted by the Ellen MacArthur Foundation, and over 350 organisations have endorsed the six key principles quoted below: [14]

1. Elimination of problematic or unnecessary packaging through redesign, innovation and new delivery models is a priority.
2. Reuse models are applied where relevant, reducing the need for single-use packaging.
3. All plastic packaging is 100 per cent reusable, recycled or compostable.
4. All plastic packaging is reused, recycled or composted in practice.
5. The use of plastics is fully decoupled from the consumption of finite resources.
6. All plastic packaging is free of hazardous chemicals, and the health, safety and rights of all people involved are respected.

The climate is changing – you can too

All in all, thinking about your plastic footprint makes you a more mindful consumer. We are so used to take, take, taking all the time, but, actually, if you take a second to ask yourself, 'Do I really need that?', or 'Do I really want that item wrapped so heavily in plastic for two seconds of pleasure?' Most likely your answer will be no. That goes hand in hand with one of the main themes of this book: becoming mindful. The plastic conversation is where the phrase 'one person can make a difference' really comes into effect.

First things first: reduce your plastic consumption The easiest way is to stop buying things you don't need, but, when necessary, look out for available options that don't come wrapped in plastics.

For those of you who feel the rage of unnecessary items packed in layers of plastics – such as fruit and vegetables wrapped at the supermarket – unwrap your packaged items and leave them with the cashier in protest of better practices. Don't worry, they won't think you're crazy; people have been doing this for a while. It does feel liberating.

Always come prepared whenever possible with your own:

- Shopping bag
- Bottle (if just one in ten Brits refilled their own reusable bottle just once a week, we'd save about 340 million plastic bottles a year).
- Coffee cup
- Cutlery

At the beginning it can seem a chore carrying so many extra items, but after time it becomes part of your routine. Using reusable items often comes with a saving, too. If you start to add up all the bottles of water you buy for £1 or more, you will save money significantly. Lots of businesses are now offering incentives for bringing your own reusable cup, with discounts of up to 40 per cent off your drinks.

Say no to drinking straws – with the exception of those people who need straws from a disability aspect. Straws can be a vital attribute to independent living, and an outright ban should take this group into consideration.

Use your social platforms to spread awareness about plastic pollution, and encourage others to reduce their consumption. Set a good example for younger generations.

Use refillables at home The easiest way to be more cost effective and eco-friendly is to buy bulk products of shampoos, conditioner and home products, and refill them accordingly.

Some supermarkets and brands such as Loop Store offer refillable dispensaries. They are, however, still fairly unusual, but keep an eye out, as this market is set to grow fast in the coming years.

Avoid microbeads These are small, solid, manufactured plastic particles that are less than 5mm in size and do not degrade or dissolve in water. They are mostly added to beauty products such as facewash and cleaning products. Microbeads are not captured by most waste-water treatment systems and so, when they get washed down the drain after use, they end up in our oceans. In 2018 the US and British governments banned the production of microbeads, but in many countries partnerships for further bans must be initiated. For manufacturers, there are several natural biodegradable alternatives to microbeads that have no environmental impact when washed down the drain; some examples include ground-up almonds, oatmeal, sea salt and coconut husks.

Call for clearer labels On many occasions, you have an item that bears the recycle icon but it turns out that you can't recycle the whole thing. Companies should be called out for clearer instructions on which parts of the wrappings can and can't be recycled.

Be your own five-minute hero, and whenever you find yourself next to a beach or river, spend five minutes picking up plastics along the shore and dispose of them correctly. #5minutebeachclean

Get involved with larger beach clean-up initiatives if you live by the coast. River and park clean-ups are also popular occurrences and great ways to socialise with like-minded people in your area.

Be aware of product wrapping labelled 'biodegradable' If a product is biodegradable, you might assume you can throw it into the environment and it will disappear, but this is mostly a myth. This book is about being as realistic as possible and although, yes, biodegradable sounds more appealing, it doesn't usually mean that the product will degrade in *any* environment. In most cases, only industrial composting under specific conditions can ensure biodegradable objects break down within a reasonable period; otherwise systems are not currently in place to recognise and process these types of materials.

Walk around your workplace and take note of where single-use plastic could be avoided, from the canteen to the coffee station, including unnecessary stationery. You are essentially campaigning for better working conditions that are, after all, basic human rights.

ONE SIMPLE CHANGE – A BIG DIFFERENCE

My non-profit organisation, the Ocean Generation Foundation, once collaborated with a well-known large corporate in London. To initiate the partnership, I had the opportunity of meeting with an executive at his office and, naturally, we began talking about an event we were going to host at that company's space. He started asking me why I cared about the topic and I highlighted how small differences can make a big impact. At that meeting we were offered single-use plastic bottles of water. I explained that by just switching these plastic bottles his company could immediately start making a positive impact on the environment. The next time I saw him he told me how he had looked into it and his floor in the building alone used over 52 full crates of plastic water bottles each week.

Shocked at the sheer number, he encouraged all his team on the entire floor to each invest in a reusable water bottle and to fill it up at the stations. Just one switch on one floor of an office building is potentially saving thousands of plastic bottles every year from entering our oceans.

I have shared this example because at work we need to have the courage to spark a conversation. That's when change really starts to happen: when we vocalise the issues. Your boss, or any CEO or manager, is a human being and, ultimately, they will want the best image for their company and the welfare of their employees.

> 'We all know about plastics, how appalling it is and what damage it does. We have been saying it for years. Yet suddenly the bell rang and everyone is aware of it. That does give one hope. It's spreading all around the world.
>
> 'But Rome wasn't built in a day.
>
> 'The movement has started and it's unstoppable now.'
>
> **Sir David Attenborough**

3

The Ugly Truth About Fashion

'Sustainability has been trending for billions of years, or we wouldn't be alive. It's excess that is the trend, and we need to make it firmly out of fashion'

Orsola de Castro, co-founder of Fashion Revolution

Not so very long ago clothes were relatively expensive for everyone. Decisions about what to buy were made carefully, and most people didn't have a vast wardrobe of clothes that might be worn only for one year, or even one season. But with the growth of fast fashion – clothes that can be bought cheaply and are not intended to last – as well as the synthetic materials they are made from, the clothes we wear have now become a subject of environmental concern.

Think about the last item of clothing you purchased.

- Did you reflect on any ethical aspect when you bought that item?
- Who made the fabric?
- Who stitched that fabric together and perhaps embellished the piece?
- How much water was needed to make the item?
- How much energy and carbon dioxide has been released in the production and transportation of such a garment?
- How will your purchase affect landfill when discarded?
- Was this garment linked to any human trafficking operations or child labour throughout the supply chain?
- Were there harmful chemicals associated with the production of this garment?

Be honest. Without feeling any guilt, did you actually consider even one of those questions before buying your last item of clothing?

Fashion out of control

The fashion industry accounts for over 10 per cent of global carbon emissions and remains the second largest polluter on the planet after oil and gas. Yes, there are disputes about exactly how accurate those hefty statistics are, but, to put it into perspective, every single day we get dressed and we eat. Just like food, fashion is an essential industry for every human, and it's an emotional decision.

The fashion industry has been unregulated from the very beginning of its time. For far too long fashion manufacturers and brands across all aspects of the supply chains have violated the rights of people and damaged the planet. To put it into context, the industry is destroying rivers and fresh-water sources; 85 per cent of plastic pollution in the ocean is due to microfibres from synthetic clothing, and these are choking our sea life.

Fashion is one of the most labour-intensive industries, directly employing at least 60 million people. But along the supply chain the industry oppresses millions of farmers and garment workers by forcing them into modern-day slavery, simply because fast fashion prioritises profit over people.

Can we change the way fashion works today?

The pleasure of shopping, the temptation when walking down the high street, the endless adverts across social media, and the rise of influencers as walking brand campaigns – all these enter our subconscious and increase our desire for more, more, more. Britons buy more new clothing than any other country in Europe, and cheap clothing is now copiously available. Many of us are slaves to fashion trends and, as new clothing comes into our lives, we also discard it at a shocking pace. The amount of clothing and textile waste we produce today globally is in the millions of millions of toxic tonnes, and very little at the supply-chain level is being done to address the magnitude of fashion pollution.

In a world increasingly motivated by the bottom line of profit, are fashion and sustainability paradoxical? What does the future of fashion really look like? It is a topic that is coming up in so many conversations; lots of new summits are organised by fashion insiders to discuss where and how the industry must lead. The oxymoron word is 'consumption'. How can the fashion industry, built on growth, respond to the fact that we already have excess clothing at home? How willing and how far are brands determined to go to make fashion fair for people and the planet? How far are we, as consumers, willing to show brands that we want sustainability?

What we believe in can be forgotten when we shop

Research and studies are showing that consumers are becoming more concerned about the impact of their clothing choices, but data shows that millennials rank everything else over sustainability, such as ease of purchase, price, uniqueness and the brand name. A disconnect exists between what clothing consumers want and what clothing we buy, however. The website Entry Level Activist has taken the psychological theory of 'cognitive dissonance' and applied it to the fashion sphere. Cognitive dissonance is the discomfort experienced by someone who holds two or more contradictory thoughts, ideas or values. If you believe that unethical fashion is wrong, but you shop from unethical brands, that's cognitive dissonance. When we behave in a way that contradicts our beliefs – environmental, socio-economical and political – as consumers we can actually begin to experience physical discomfort and mental stress. These feelings can be heightened when we realise that we really do have a choice to act in alignment with our beliefs, but what is most difficult is aligning our individual actions with the systematic processes we live within.[1]

Ultimately the fashion system is a product of the global economic system based on consistent growth. Consumption is at the core, and the idea of asking consumers to interfere with the system becomes a very abstract request. The industry requires systemic change, and the fashion industry as a whole needs to take responsibility for creating our dangerous obsession with fast fashion. When exploring the fashion industry there are many overlaps between economic, environmental and social justice. This chapter might be filled with many more questions than answers, but keep in mind that it's not just about saving the planet, it's about saving the people who make our clothes, too.

The hard facts: fashion

- Eighty billion pieces of clothing are consumed globally every year. This is 400 per cent more than the amount we consumed just two decades ago.[2]
- The fashion industry accounts for over 10 per cent of global carbon emissions and remains the second largest industrial polluter, second only to oil.[3]
- Nearly 70 million barrels of oil are used each year to make the world's polyester fibre, which is now the most commonly used fibre in our clothing and takes hundreds of years to decompose.[4]

- A study by the Ellen MacArthur Foundation found that the equivalent of one rubbish truck of textiles is wasted every *second*.
- Eighty-five per cent of plastic pollution in the ocean is due to microfibres from synthetic clothing, threatening marine wildlife and ending up in our food supply.
- Over 65 million trees are logged every year and turned into fabrics such as rayon, viscose, modal and lyocell.
- A quarter of the chemicals produced in the world are used in textiles.
- A single T-shirt takes 2,700 litres of water to make – the same amount of water an average person drinks over the course of 900 days.[5]
- One in six people work in the global fashion industry.[6]

Fast fashion

The contemporary term 'fast fashion' is used by fashion retailers to express that designs move quickly from the catwalk to capture current fashion trends.[7]

Let's be real. Fast fashion completely distorts our sense of value. A coffee in London can easily set you back £3.40 – this is a caffeinated drink that you will most likely consume in less than 30 minutes.

You can buy a T-shirt from a high-street retailer for £1.50. A T-shirt that has taken over 2,700 litres of water to make, touched by approximately 55 people in the supply chain, from picking the cotton to sewing the garment, and it has travelled across the world creating a huge carbon footprint on its way to you. How is it possible that a coffee is more expensive than a T-shirt?

In the past, clothing was something practical. Something we valued, that got passed down between generations, fixed when it was broken, and it was only thrown away when it was genuinely unfixable. When we discuss fashion through a sustainable lens we tend to lament with nostalgia our parents' generation and 'how it used to be': passing clothing from one family member to another, fixing it until the very end of its potential life. But, to be honest, using the generational argument as a sustainable case is not realistic in the 20th century. People just didn't have nearly as many clothes previously, especially when young – mostly because they couldn't afford it. We have gone to the other extreme today. Price is the key and a huge reason as to why we are where we are today; if clothes are cheap and affordable, they will be bought. It's not such a sexy story anymore when we have thousands of high-street and online retailers winning off the back of this disposable culture we have created. The fashion system indulges consumption, enables certain lifestyles and opportunities.

The short lives of today's garments

Fast fashion garments tend to be worn less than five times and kept for 35 days before they are thrown away. British youth tend to classify a garment that they have worn just once or twice as 'old'. Fast fashion companies design clothes that intentionally fall apart quickly. They pursue a strategy called 'planned obsolescence'. This means that they design garments to become unfashionable, wear out, lose shape or fall to pieces easily to force consumers to keep buying new clothes. We know this because once upon a time there were just two fashion seasons: spring/summer and autumn/winter. Fast forward to 2014 onwards and the fashion industry is churning out up to 52 micro seasons of clothing per year. These collections are so cheap that they allow the consumer to feel rich because they can buy so much.

The hidden story of fashion

The production and consumption of fast fashion does not come without consequences. The fast fashion movement and its 80 billion new garments created every year are polluting our planet at an unprecedented rate and prohibiting millions of people, especially in the developing world, from earning a fair wage and living with dignity.

Western countries are generally far away from clothing manufacturing sites. Over 75 per cent of global clothing exports and 50 per cent of the world's textile exports are produced in developing countries.[8] The reality of the environmental impacts is far removed from people like us, and it develops a disconnect between the responsibilities of our own consumption and waste.

We are often blind to the over 92 million tons of solid waste dumped into landfills each year. Through the wasted textiles we throw away, or the millions of garments we send to landfill that get burned, we waste valuable resources: the 1.2 billion tonnes of CO2 emitted annually throughout fashion supply chains; and the polluted waterways and the sweatshops where manual workers are employed for their hard labour under very poor working conditions at extremely low wages, including child labour.

Don't be fooled, either, that it's only high-street brands that are part of the unethical labour and horrendous environmental practices. Fast fashion is often just a scapegoat; luxury brands are notorious for utilising worse practices than the lower-end brands. A lot of high-end luxury designers put their labels on cheaply made clothes manufactured in the developing world.

Thus, fashion for a couple of decades has caused severe harm, from environmental degradation to human rights violations.

The rise of online shopping

No one ever thought that big high-street brands might ever be in peril, but, due to the rise of the Internet, more and more physical stores are closing, and more of us are buying with a click of a finger.

This transition from physical buying to online could have provided a window of opportunity for brands to do the right thing and take sustainability into account, but unfortunately that didn't happen. Instead, it made way for online retailers to recreate the bodycon dress-look worn by an influencer less than a day ago and sell it to the online masses. We are at a point where both the industry and ourselves as consumers recognise that things need to change, so let us delve deeper into the industry.

THE MOST COMMONLY USED FASHION TEXTILES

Polyester is the most commonly used fibre in our clothing, and it is essentially *plastic*. Polyester is a shortened name for a synthetic, human-made polymer, which is most commonly referred to as polyethylene terephthalate (PET), made by mixing ethylene glycol and terephthalic acid[9] – essentially a form of crude oil. According to a *Forbes* study, 70 million barrels of oil are used each year to make the world's polyester fibre. It takes more than 200 years to decompose.[10]

Cotton Ninety per cent of cotton is now genetically modified (GM) and is the world's single largest pesticide-consuming crop, which uses vast amounts of water. The impacts of these chemicals (herbicides, insecticides, fungicides, and so on) on the land and on human health are largely untested, but they are dangerous for those working in the industry. The chemicals from GM cotton are passed into the bloodstream via the skin, so people who work in the cotton industry are at risk of chemical exposure, but so is anyone who wears garments made with this type of cotton. To avoid this, try looking for and buying *organic* cotton wherever possible.

▶

Leather has been popular since 3000 BC, but leather is not really a by-product of food production and is not luxurious. According to designers such as Stella McCartney, who do not use any animal products, leather is linked to a variety of environmental and human health threats. The amount of feed, land, water and fossil fuels used to raise livestock for leather production come at a huge cost to our planet. Apart from raising the livestock needed, the leather-tanning process is among the most toxic in all the fashion supply chain. Workers are exposed to harmful chemicals, while the waste generated can pollute natural water sources, leading to an increase in diseases such as cancer in surrounding areas.[11]

So what are the alternatives you may ask? It's not cool to be cruel but the alternative faux leathers come at a significant environmental cost, too. Both polyurethane and polyvinyl chloride compounds must undergo chemical processes to make them flexible enough to imitate leather. This requires toxic solvents derived from fossil fuels, which we know take hundreds of years to biodegrade or release a tonne of harmful chemicals if burnt at landfill. This hasn't stopped brands releasing statements screaming 'sustainable leather alternatives', despite the downsides.

So, ultimately, the choice is down to you as the consumer. Is it better to buy real leather, as you will likely care for a hand-crafted real leather item and may even pass it down, as opposed to the cheap £5 pleather option? There are always

▶

trade-offs to consider. Designers are currently looking into lab-grown alternatives, such as pineapple, fungi and other vegetable leathers, but the research is in such early stages it's hard for it to become scalable right now. So, at the moment, the best alternative is using what already exists, which is the deadstock or by-products of the meat industry, ensuring nothing goes to waste (brands like Gabriela Hearst are great examples of minimising waste).

Wool is a very traditional fibre but wool has a tiny and decreasing share of the world textile market at around just 1 per cent. About 2 billion kilograms of raw wool per year are produced from a global herd of around 1.2 billion sheep, each sheep providing around 4.5kg of wool per year. That roughly equates to one wool sweater per person per year for everyone on the planet! Other types of wool include cashmere and mohair, which come from goats, and alpaca for a similar fleecy material. Wool is another divisive fibre, however. Due to decreasing demand and the increased use of synthetic fibres, in the 1970s the search for a wool technology that could make it easily washable and even withstand tumble-drying led to the production of wool treated through acid baths and other more toxic processes.

Advocates say wool is an all-natural, renewable fibre and tends to be washed less frequently than other fabrics, while offering better insulation than other materials. Naturally however, wool is derived from sheep, and animal welfare organisations and concerned citizens are committed to the well-being of animals during this process, even calling it 'punching and stomping' on sheep. Secondly, some environmentalists have pointed out that sheep used for wool contribute to climate change by threatening land, air and water sources.

Silk For hundreds of years silk was essential for a huge range of fashion items and it is often associated with the height of luxury. The delicate fibre is an ancient, natural protein fibre made from the thread of silkworms, which can be woven into textiles.

It is a highly renewable resource and biodegradable at the end of its life. But as silk has become commercial, the silkworms that once just fed on mulberry leaves, now require the use of pesticides or fertilisers to grow. Producing a small amount of silk requires vast resources, and it can also be a very labour-intensive process. There is also a cruelty element to most silk production as the worms are boiled alive inside their cocoons. If you are buying silk from a big brand it most likely is coming from an industrial facility in China, but there is such a thing as 'peace silk', a type that allows the moth to emerge before harvesting the cocoon, and is manufactured in the most stringent social and environmental standards, mostly in India. Silk is a beautiful fibre and, if you look after it, it will last for many years, with the substitute being rayon or polyester.

Microfibres

We talked about microplastics in the previous chapter, but it's microfibres in our clothing that are the most widespread and lead to environmental damage.

As we have seen, the most commonly used materials for clothing today are polyester or nylon, which is essentially plastic. Our clothes are made up of tiny plastic fibres, which are so thin that they are practically invisible to the human eye, and these are shed every time we wear, wash or dispose of our clothing. A single washing machine load can release 700,000 fibres to waste water, according to research from the University of Plymouth.

The microfibres are eventually discharged into our oceans through the waste water from our washing machines. Here, they are easily mistaken as food by those at the bottom of the food chain, such as plankton. The microfibres are eaten by the plankton, which are then eaten by small fish and then larger fish, and in this way they are passed up through the food chain to much larger fish and eventually to our own bodies if we eat fish. While most plastic has been found in the guts of fish, and would therefore be removed before eating, some studies have proven that microplastics, particularly on the nanoscale, do transfer from the guts to the meat (and we do eat some shellfish and small fish whole). In 2011 in the river Clyde, in Scotland, 83 per cent of Dublin Bay prawns, the tails of which are used in scampi, were found to have ingested microplastics; so had 63 per cent of brown shrimp tested across the Channel and southern part of the North Sea.

To stop microfibres from entering the oceans you can buy

Guppy Bags. These are mesh-like bags that you put your clothes inside before placing the bag into the washing machine. The bags capture thousands of microfibres when you wash your clothing, allowing you to dispose of the fibres responsibly in the bin.

> 'I get frustrated that 90 per cent of environmental issues mentioned in fashion are marketing'
>
> Stella McCartney

Is the fashion industry changing?

For decades we have seen the same brands dominate the fashion markets, and for the same amount of time representatives have woefully spoken about a 'desire' for change, with better collaborations, while simultaneously using the 'eco' word as an opportunity for a great PR story. But sustainability has been on the global agenda for such a long time that the pace at which the fashion industry has committed to helping people and the planet is simply not good enough. The fashion industry needs accountability and leadership by mobilising the system to change the way it produces, markets and consumes fashion. Responsibility must be taken by leading sustainability experts, fashion decision makers, CEOs and influencers to showcase innovations that would make the world a better place.

Individually, bigger companies are generally pro-active and

intelligent on sustainability, but collectively they are not coming together. The private sector is not used to working together because the natural market instinct has been to compete; however, it can't just be a few big brands leading the way for more ethical supply chains or the small start-up brands to create with purpose. Instead, the sustainability fight must be an industry-wide collaboration. The fashion industry needs collective intelligence without upsetting the world economy.

Brands have the power to drive positive change through researching and developing new and forward-looking textiles, systems and services that adhere to more responsible production. If we look at sustainability from a designer's perspective, the creators are the ones that determine what is cool. Designers are crafting great products, so let us take away the 'sustainable' terminology from the very start and stop making it the exception but the actual design norm. Supply can be the driver of demand if the supply chain changes.

In response to the younger generation's passion for environmental and social causes, brands have become more purpose driven to attract consumers. From launching conscious campaigns and capsule clothing, larger brands recognise the power of purpose, but how far are they willing to go to implement change?

Greenwashing – insincere environmentalism

Greenwashing is the practice of making an unsubstantiated or misleading claim about the environmental benefits of a product.

In the fashion world, greenwashing has become a common practice, with deceptive PR and marketing that make unsubstantiated claims about the environmental benefits of garments and accessories, in order to promote the perception that these brands are environmentally friendly. There are many companies who are grossly overstating environmental and ethical benefits of new clothing and accessory capsules. Just because a company improves its 'guidelines' around the topic of sustainability, doesn't mean it has actually implemented any change across its supply chains. Details as to how transparent these brands are is key – it is not enough to say they are 'conscious'. If the brand is vague about the source of its materials, its workers or any other social commitments, then it's most likely something or someone was exploited in making its products. Be curious and do your research as a consumer.

The consequences of greenwashing are not only that it misinforms consumers about what they are buying, but it also creates an unfair and uneven playing field for those brands that are investing in accountable and transparent supply chains. These are usually the small brands that are taking on the world challenges and, of course, incurring higher costs in the name of ethics.

Myth buster: clothes made from recycled plastic are good for the environment

We are seeing a surge of clothing made from recycled plastics, such as sports gear and swimwear. It is amazing to see brands taking the initiative, but it is important to highlight the flip side. When we talk about the invisible threat of microplastics washing into the seas and affecting the food chain, as I have explained above, recycled plastic clothing is contributing further to this issue, not ameliorating it.

Clothes made of recycled plastic are just filling a void, not solving the problem. Although recycling plastic into clothing is a good effort in promoting the circular economy, it is by no means the solution. In fact, we should be asking how can we eliminate the need for plastics in our clothing altogether?

New textiles for a new ethos

Sustainable textiles can be defined as those that don't exhaust renewable resources, and, thanks to modern technology and science, there are designers around the world who are developing materials that don't pollute our environment during or post manufacture. Below are some of the viable alternatives to fashion industry norms, which need help to become mainstream:

Pineapple fabric

London-based Ananas Anam has developed a vegan alternative to leather made from pineapple leaves, known as Piñatex. The pineapple leaves used are a by-product of the pineapple harvest in the Philippines that are then processed by a method known as 'decortication', which essentially extracts the useful fibres from the leaves. The fibres then undergo an industrial process to become a non-woven textile that is strong yet versatile and soft.

To add to the fundamental Piñatex values concerning the environment, the by-product manufactured in the process is biomass, which is converted into organic fertiliser or bio-gas and used by farming communities, thereby closing the loop of the material's production and promoting economic and social sustainability of the neighbouring villages.

Coffee-ground fibres

Most coffee drinkers at home, and the larger coffee chains, simply throw away coffee grounds after their morning coffee. But it has been discovered by a textile firm in Taiwan that used coffee grounds are an important raw material from which to make coffee fibre. Through a technology that is normally used to turn bamboo into a viscose material, coffee grounds with a small quantity of polymer elements, can be spun into yarn for a variety of products, from outdoor to home wear. The coffee yarn offers excellent anti-odour qualities and a quick drying time.

Chitosan or chitin fibre (essentially, crustacean shells)

Designed by manufacturers in Germany, and probably one of the most memorable alternatives, chitin is related to cellulose but is a hard, semi-transparent substance that makes up the outer covering of crustacean shells such as crabs and lobsters. It is mixed with viscose to create fabrics that are hypoallergenic and biodegradable.

Salmon leather

In the fishy category also falls salmon leather. Made out of salmon skin, which is a by-product of the fish processing industry, it is therefore a material that would otherwise be discarded. It is tough and resilient with a smell that reminds you of the sea (don't worry, it's not fishy, it's more of a light, earthy smell), perfect for shoes and accessories.

What will make us change?

There is a gap between what consumers are saying and how they are behaving. Consumers *say* they want more sustainable clothing, but they are still *buying*, mostly, unethical fast fashion. If sustainability is really in, the hardest question is why aren't more people buying ethically made clothes?

Part of the answer is a lack of availability, supply and convenience. Consumers desiring more sustainable goods find that they are just not sufficiently readily available on the market. In addition, money and the price of sustainable clothing is often a deciding factor for consumers and can interfere with people's desire to do the right thing. Whether consumers are not willing, or not able, to pay more for sustainable clothing and thus opt for fast fashion that gives them 'value for money' is the question. The idea that the consumer should pay more for sustainable products is an absolute no-go for brands, who believe that the consumer that has evolved over the last ten years is the most informed, but that it should not be their burden to pay more for ethics. As noted by the director of sustainable business at Marks & Spencer, at the Copenhagen Sustainable Fashion Summit 2019: 'It must be the responsibility of the brands who have the relationships with the hidden supply chains to solve the problems.' This will only start when brands admit that they have got rich from exploitation but commit to accountability and transparency moving forward.

The flip side is if consumers understand what a better investment might be, they could almost be guaranteed to invest in timeless, quality clothing that will last much longer and have a positive effect on the environment. The genuine challenge is while we wait for brands to make sustainability the norm, there are many things that we, as consumers, can do to make more sustainable choices. If the brands and corporations won't move faster, we must ask ourselves, 'Do we care?' Only you can answer that, but if you *do* care, let us ignite the fourth revolution, a combination of technologies changing the way we live, work and interact. The common theme among each of the industrial revolutions before us was the invention of a technology that changed society fundamentally. The First Industrial Revolution in Britain was powered by the invention of the steam engine, the Second was roughly characterised by mass industries like steel, oil and electricity, while the Third is said to revolve around the digital sphere as a result of the creation of personal computers and the internet. The Fourth Revolution represents an opportunity – enabled by technology – to recognise and harness the new in order to create an inclusive change. You can start igniting the Fourth Revolution by asserting your voting power through spending your pounds and consuming as consciously as possible.

Renting – the end of ownership?

Across lots of categories consumers have demonstrated an appetite to shift away from traditional ownership to newer ways in which to access a product. Since the rise of car-sharing platforms such as Uber, home rental through Airbnb, and Netflix replacing video stores, there is a fundamental shift in consumer behaviour, and it is expected to dramatically impact the fashion business in the near future. Clothing rental platforms are shifting towards a sharing economy. Rental also encourages consumers to read descriptions, look at size charts and to consider their purchases more carefully, instead of fostering this buy-now-return-for-free culture, which has a huge carbon footprint.

Renting clothing used to be a niche business used only for occasional wear. Today renting services are disrupting the fashion space at an unprecedented rate with the aim of extending a garment's life – especially luxury ones. One in seven consider it a fashion faux-pas to be photographed in the same outfit twice.

This model does have to evolve however, as currently demand outstrips supply, and it is much easier to rent if you are a smaller size living in a major city, especially in the US and Europe, but in China, YCloset takes a different approach. It uses a subscription rental model to grant customers access to an array of clothing and accessories included in the fee. If a customer decides they like a particular piece they also have the option to buy it outright. Whatever the rental model, over the next decade, from a sustainability angle, it proves there is absolutely no reason to fill up our landfills every single year with discarded clothes.

> 'The future of fashion will be rooted in accountability'
>
> **Wilson Oryema**

What is the future for fashion?

This book can't predict what will happen with the future of fashion, but here is an outline of what the right direction could look like for brands who want a more sustainable fashion industry:

- Take responsibility and account for the entire supply chain. That starts with the raw materials grown to make the fabrics in the farming sector, the production of textiles and, lastly, better labour protection for garment workers, no matter where they are in the world.
- Use the Higg Index: a new tool created by the Sustainable

Apparel Coalition and launched to standardise how apparel companies can measure their environmental impact through their supply chains.

- Choose more responsible methods of production, from sourcing organic cotton grown without pesticides to dyeing fabrics and leather with vegetable tan instead of toxic chemicals.
- Make supply chains transparent for consumers. Allow buyers to access information so that they can follow the journey of how their clothes are made. Our ignorance of the reality of what goes into making our clothes is also a reason why we buy more, because we don't stop to think.
- Online retailers need to be more mindful of their packaging and carbon footprint. Stop packing oversized boxes with useless paper and waste. Rethink the size and design of packaging along with its material. Incredible innovations, from sugar-cane envelopes to biodegradable boxes, already exist.
- Stop creating up to 52 seasons a year. Instead, stick to the basic four seasons that humanity has always known: autumn, winter, spring and summer.
- Designers can set the tone for what is cool by designing more timeless pieces.
- Get innovative with dead stock. The media exposed how brands were burning new clothing items. Instead, re-purpose, re-use, share, discount new clothing – for goodness sake don't landfill new garments. In the current climate of tense politics and uncertain trade conditions,

companies could seek new opportunities to sell dead stock outside their traditional markets in the UK and Europe.

- Brands making profit can give back to society, support charity organisations and environmental initiatives.

Switching to more sustainable practices might hurt the bottom line at the very beginning of any transition, but, hey, there will be nothing to buy if there is no planet.

Fashion heroes

Having realised the role of fashion brands in protecting and optimising our planet's finite resources, there are certain leaders who are inspiring their own companies to integrate sustainability into their entire business model. Right now, there are the fashion heroes who recognise that values and social purpose must be included in company ethics. At the other end of the spectrum, however, are the fashion companies that decide to sit back and wait to see if there is a business case for sustainability before implementing any progressive change. This is creating an unequal playing field for the businesses leading the way, but those who do are attracting the best talent in the industry and forging new relationships with their customers. Of course, established brands and high-end designers can, to a degree, charge accordingly for any change in practices, because of their naturally higher price-points, while high-street brands often insist that it is low pricing for the masses that gives them their competitive advantage. But many of these are multi-million-pound corporations that could easily use some of their profits to investigate their third-party supply chains, ensure transparency and begin identifying the steps needed to move towards more sustainability. However, this will only really change when consumers start pressuring the high-street brands to act in accordance with the way the world is moving.

Lastly there are start-up fashion hero brands that start with sustainability and purpose in mind such as VEJA, Reformation or PANGAIA. Brands like these refrained from producing or selling until they had their supply chains figured out.

Here are some of the iconic designers who have stepped up to the plate and are leading the fashion industry towards a greener future:

Katharine Hamnett

There is nobody better versed than fashion's original eco-warrior to tell it how it is in the fashion industry. Starting with a very glamorous trajectory in the fashion world, with a brand that was once one of the largest retailers, Katharine Hamnett had an epiphany on what was right. Ditching the glamour, she has been on a decades-long journey to make not only her own brand sustainable but to disrupt the entire fashion industry into being more accountable for its actions.

Stella McCartney

A lifelong vegetarian and vocal animal rights activist, Stella decided at the very start of creating her own label never to use leather, fur or feathers in her designs. Since the 1990s, she has bioengineered spider silk, mushroom leather and recycled ocean plastics, and never, ever compromised. She and her brand are what the future of sustainable companies looks like.

Sébastien Kopp and François-Ghislain Morillion

Co-founders of VEJA, these two guys are taking the sneaker market by storm. Look down on the streets and you will notice the iconic sneakers with a V for VEJA on all types of feet. Why? These guys have struck a chord with audiences, not just because they focus on transparency, environmental responsibility, fair trade and social justice throughout their supply chains, but also because their sneakers have slick designs. Every step of the sneaker-making process is unique; from wild rubber from Amazonia for the soles, organic fair-trade cotton, upcycled plastic bottles, upcycled fish leather and new fabrics made out of corn waste. As a result, they have built a multi-million dollar business proving that style and ethics are a killer combination.

Livia Firth

Livia, along with her brother Nicola Giuggioli, founded Eco-Age, a UK-based consulting company that helps brands develop ethical and sustainable supply chains. She has serious eco-warrior credentials and has convinced big-name celebrities to join her Green Carpet Challenge, encouraging stars to wear eco-dresses on red carpets worldwide and shining a bright light on the true cost of fashion.

Vivienne Westwood

'Buy less, choose well, make it last' are Vivienne Westwood's famous words. Throughout her long career, the punk-rock designer has continuously presented protest collections with strong climate messages. Her catwalks resemble protests, and she has collaborated with Greenpeace on numerous occasions and published *Get A Life: The Diaries of Vivienne Westwood*, a book in which she discusses her solutions to the world's polemics. Vivienne is a true visionary.

What we can learn from these fashion heroes is that nothing is ever perfect; it's about starting somewhere and riding the journey together.

The climate is changing – you can too

As with purchasing the products of other industries covered in this book, your money is power. Here are some of the things that *you* can do to have a more sustainable wardrobe:

The ultimate solution is for you as the consumer to buy better, buy less and care more.

Buying better means being conscious of what we buy from the very beginning of building our wardrobe capsules.

Choose quality and timelessness over cheap trends.

Cultivate your wardrobe with local designers wherever possible.

Vintage fashion is proof that a garment can last many years if it was produced with care from quality materials, and it can still provide you with the emotional pleasure of getting a good bargain.

Look at the labels on clothing and where possible opt for fewer synthetic fibres and more natural materials. Favour the use of non-animal derived materials. Look out for plant-based leather alternatives such as pineapple, mushroom and algae.

Take the 30 Wears Challenge: a global movement invented by Livia Firth's Eco-Age. They suggest asking yourself, 'Will I wear it a minimum of 30 times?' If the answer is yes, buy it. But if not, do not purchase that item. You will be surprised at how many times you say no to yourself.

Reduce the number of times you wash your clothes. Obviously wash your underwear and any tight-fitting clothing that might be soaking up body odours. But the reality is that most garments do not need to be washed every single time. Jeans, for example: manufacturers advise that they should not be washed after every use.

Wash smarter Make sure you only switch on the machine when you basically have a full load; there is no need to waste so many litres of water for one garment. Wash at reduced temperatures, preferably 30°C or less. Don't forget to use your Guppy Bags (see page 86) when washing, too, to prevent microfibres entering our waterways and oceans.

Choose your fabrics with thought Almost all clothes you buy come from four sources: plant, tree, animal, oil (or a combination of them). What you want to look for are natural organic fabrics such as cotton, hemp, linen, and so on. Organic means less water and fewer pesticides have been used and production conditions are safer for workers.

Recycle your old clothes Less than 1 per cent of clothing gets upcycled into something new, so being smarter about recycling our old garments can make a massive difference. Donate to local charity shops, but remember that everything must be in a clean, wearable condition. Ask yourself if would you consider buying that item at a charity shop. If the answer is no, you need to dispose of your items via your local authority recycling centre. It is probably best to check with your local council on what policies they have.

Look out for local businesses and initiatives that promote circularity, such as take-back schemes, repairing and recycling solutions, or upcycle and repurpose garments.

Shift away from the traditional ownership models and rent your clothes. Do you have a special event to attend or do you easily get bored of your wardrobe? Then start renting. Rent your clothing from websites such as: Rent the Runway, HURR Collective, Front Row.

Don't be fooled by high-street fashion brands jumping on the sustainability trend. They might look like they are saying all the right things to tick the right boxes, but they can be ignoring key issues and not actually practising sustainable values. Do your research, and if the campaigns look wishy-washy, there are plenty of amazing sustainable brands out there working with local communities and producing ethical clothes for you to buy.

Read some books to get you mad about fast fashion:

- *To Die For: Is Fashion Wearing Out the World?* by Lucy Siegle
- *Wardrobe Crisis: How We Went from Sunday Best to Fast Fashion* by Clare Press
- *Fashionopolis: The Price of Fast Fashion and the Future of Clothes* by Dana Thomas

Educate yourself further by watching fashion documentaries such as:

- *The True Cost*
- *Stacey Dooley Investigates: Fashion's Dirty Secrets*
- *RiverBlue*
- *Alex James: Slowing Down Fast Fashion*
- *China Blue*
- *The Next Black*

4

Every Bite Counts

'The biggest opportunity to fix the food system: consumers knowing and caring about how their food is produced'

Bruce Goldstein, Farmworker Justice

Go vegan. End of chapter.

Veganism is perhaps the argument that most would expect from a book like this. I would like to emphasise that from an animal welfare perspective going vegan is the fairest option for our planet. The rest of the arguments and the research into our food systems not only relate to climate, but to people and economies because they are intertwined and require much broader discussion. The Earth is an extraordinary complex eco-system and any one-size-fits-all solution risks failure.

Food is one of the most important industries to humanity: every living thing on the planet depends on it. Food is also part of the emotional fabric of our lives: brunch on Sundays, dinner with family and friends, cooking – all are reasons to socialise and celebrate. We all share the one thing that can have a significant impact on climate change: the foods we eat. But what we are consuming represents a global health and environmental crisis happening today. Why? Because our food production and consumption habits have become major drivers of climate change, water stress, land grabs, biodiversity loss, soil erosion, deforestation and the depletion of fish stocks. It has become apparent that our planet cannot sustain our existing food habits, and we are going to have to shift fundamentally – and quickly.

What sort of diet could work for you?

Most environmentalists will tell you that one of the greatest ways to reduce your carbon footprint and help the planet is to go vegan, but you might find this difficult. If you are tempted to skip this chapter because vegans (or even the word itself) annoy you, don't worry, its purpose is not to convert you. Yes, I will discuss veganism in detail, and from an animal-welfare perspective, it is kind to be vegan and there is a cost in terms of the land used for animal farming and their incredibly harmful emissions, but it's important to consider that millions of people depend on the industry for their livelihood or simply eat meat because of geography and lack of alternatives. It is not realistic to expect the entire world to adopt veganism immediately in the name of the environment either, but how about being vegetarian? Flexitarian? Pescatarian? The list goes on. And what do they even mean?

THE DIFFERENT TYPES OF DIET INCLUDE:

Omnivore Someone who eats meat, fish and dairy produce as well as plant foods.

Flexitarian Someone who has a primarily vegetarian diet but occasionally eats meat or fish.

Pescatarian Someone who eats a plant-based diet as well as fish, and also includes dairy, eggs, honey, and other animal-derived foods that do not require an animal to be killed.

Vegetarian Someone who eats a plant-based diet and also includes dairy, eggs, honey, and other animal-derived foods that do not require an animal to be killed.

Vegan Someone who eats a plant-based diet with no foods whatsoever of animal origin (such as dairy, eggs, honey).

The bigger picture about food

Trends come and go with the latest health craze and influencer endorsement, but the conversation about climate and food goes beyond eating meat and animal welfare.

Reassessing your diet is one of the easiest ways to help the planet. The question of what we should eat to reduce the devastating impact we are having on the environment, while at the same time reducing diet-related diseases and supporting livelihoods, should dominate the food conversation.

By making small adjustments to your daily food choices, you can help respond to the climate crisis by starting with what's on your plate and making every bite count. Excitingly, though, this chapter is one of the more optimistic parts of the book, because the food sector is succeeding – more so than many other industries – in making organic locally sourced products seem aspirational and bound up with an underlying message of self-care.

To assess your relationship with food, begin by considering these two very basic questions:

1. Do you know where your food comes from (apart from your local supermarket shelf)?
2. Do you understand the environmental consequences of what you eat and drink?

The hard facts: food

- It takes on average 272 litres of water to produce a single avocado.
- Twenty-four million slices of bread are wasted every day in the UK.
- Pumpkin waste is at scary heights during Hallowe'en in the UK. Some 18,000 tonnes of pumpkin are thrown in the bin by households: that is equivalent to 360 million portions of pumpkin pie.
- Nearly 80 per cent of the world's farmland is dedicated to rearing animals.
- Livestock production is responsible for 18 per cent of global greenhouse gas emissions, according to the United Nations Food and Agriculture Organization.
- If food waste were a country, it would be the third largest emitter of greenhouse gases after China and the US.
- Over one-third of all food produced goes to waste, and the annual value of food wasted is approximately $1 trillion.
- The world loses $400 billion of food before it reaches the supermarkets, due to the lack of cold storage infrastructure, travel routes and multiple inefficient factors throughout the supply chains.
- An area larger than China is used to grow food that never gets eaten every day. It takes 25 per cent of the world's fresh water supply to grow the unused food.
- About 80 per cent of the world's poorest people live in rural areas and rely largely on agriculture. Agriculture is one of

the most powerful tools for raising the incomes and lifting people, especially women, out of poverty.

- There are over 821 million people in the world who go hungry every day. One in nine go to bed on an empty stomach. One in three suffer from malnutrition. In a world where we produce enough food to feed everyone, eradicating hunger is one of the greatest challenges of our time. We could feed all of them on less than one quarter of the food that is wasted in the US, UK and Europe alone.[1]

'A vegan diet is probably the single biggest way to reduce your impact on planet Earth, not just greenhouse gases but global acidification, eutrophication, land use and water use'

Joseph Poore

Veganism

Let's start with the buzz word 'veganism'. The single biggest message from all food research is that various pieces of scientific evidence demonstrate that our diets are one of the easiest ways to make an effective difference to climate change. By adopting more plant-based variations to your diet, you, as an individual, can dramatically save on greenhouse gas emissions, water usage and an array of other essential environmental factors. So let's look at what lies behind the typical Western diet, and its effect on the environment.

From the 20th century onwards producers realised they could keep animals such as chickens, pigs and, more recently, cattle, inside, feed them grain and they would get fatter much quicker. Animal farming has had dire consequences for the environment. The industry not only began ramping up the level of animal suffering by moving away from historic methods of livestock farming, but significantly increased its dependency on natural resources.

The production of meat on an industrial scale is highly inefficient because raising animals for food requires massive amounts of land, food, energy and water. It is hard to separate statistics and measurements that look at agricultural impacts, without distinguishing between arable versus livestock, or industrial versus small-scale farming, but either way the numbers relating to the environmental impact of our current farming methods are truly horrifying. For example, producing one kilogram of beef requires 25 kilograms of grain to feed the animal and approximately 15,000 litres of water. If all the grain used for animals was fed to humans instead, we could feed an extra 3.5 billion people.[2]

Nowhere is the environmental impact from an emissions perspective more acute than with livestock. The UN says that farmed livestock accounts for 18 per cent of all human-made greenhouse gas emissions from both the animals and their manure. To put that into perspective, that is roughly equivalent to the exhaust emissions of every car, train, ship and aircraft on the planet.

Cows are the primary issue; they release methane through their 'eruptions' – essentially their farts and burps – and in their manure. Methane is at least 28 times as destructive as carbon dioxide when it comes to heating the atmosphere.[3] Although there is significant evidence that high emissions, particularly of CO2 and methane, are associated with meat production, in the West, people still love meat, and over time it has evolved from becoming a luxury to being eaten up to three times a day at breakfast, lunch and dinner.

The magnitude of the problem can also be seen when we look at land use: 30 per cent of the planet's terrain is dedicated to livestock farming. In order to obtain the vast amount of land necessary for animal farming the process has required deforestation, illegal land grabs and loss of biodiversity because not only do we require space for animals to live or graze, as most livestock does not live off natural pastures, i.e. grass grazing, they instead rely on a diet of grains that must be grown elsewhere. The land used to grow the grains fed to animals is grossly wasteful because we could be using that land to grow grains or crops to feed people. That would also be fairer economically because feeding grain to livestock increases the global demand and drives up the price, thus making it harder for the world's poor to feed themselves.

The vegan science

One key study stands out in favour of veganism. Conducted by Joseph Poore from Oxford University, this comprehensive food study created a huge dataset that is based on 40,000 farms in

119 countries and covers 40 food products, representing 90 per cent of the foods we eat. It assessed the full impact of these foods, from farm to fork, encompassing land use, greenhouse gas emissions, freshwater use, water pollution and air pollution. Examined from a global perspective, factoring in population growth and increasing affluence in some developing countries, the figures are shocking. The study claims that livestock production accounts for 83 per cent of global farmland and produces 58 per cent of greenhouse gas emissions from food, but only produces 18 per cent of our food calories and 37 per cent of protein.[4]

There are two sides to the debate

It's no surprise that when Poore's study was published, journalists and every eco influencer jumped on the vegan message, with little broader consideration of the other issues, such as food-system sustainability. The vegan debate to some extent has been limited and dominated by those with very little experience of food production and what constitutes food security, so it is important to look at both sides of the climate argument (again emphasising that while from an animal welfare perspective it is not humane to eat any animal or produce derived from animals, the climate argument requires a broader discussion).

Poore's study suggested that we can cut global greenhouse gases by up to 23 per cent with a vegan diet alone. So a vegan diet is probably the single biggest way to reduce your impact on planet Earth, and has a far bigger effect than cutting down on flights or buying an electric car, as these cut only greenhouse gas emissions.

But going vegan is not for everyone. Many of us living in the Western world have been conditioned through societal habits to assume the source of protein with every meal must come from animals. In Britain, from the 1950s onwards, 'meat and two veg' was the staple diet for most, and every family provider was programmed to think that meat was necessary in every meal in order for it to be hearty and nutritious. Clearly this is simply not true, and over the last decade, the message from many dietitians has been that a diet of legumes (beans, peas and lentils) and various vegetables can have equivalent levels of nutrition. But we need more education on food choices and a shift in mindset for most of the population to make food switches.

Conversely, in developing countries, and for the world's poor, where people do not have alternative means, eating animals is simply a matter of survival. Research is mostly silent on aspects like the nutritional value of animal products for children in rural Africa, or the benefits of livestock areas in sub-Saharan Africa where the land is too arid to grow many crops.

VEGANISM – THE PERFECT DIET?

Without a shadow of doubt the global vegan movement is not only opening our eyes to the environmental impact of

▶

eating meat, but is shedding light on the impact of a vegan diet on our health.

Essentially, the health of the general population is turning to shit. America, especially, is becoming the home of the sick and obese. High meat consumption, especially red and processed meat, is linked with various diseases like cancer, heart disease and diabetes. In places like the US, industrial livestock farming also relies heavily on antibiotics, to control infection, thus making humans more resistant to certain medicines themselves.

If something will benefit us personally, then naturally we will be more inclined to act, and the consensus of doctors is that a vegan diet can be a healthy diet. But it is not the *only* diet, and those following a fully vegan diet must understand how to fully nourish their body and ensure they are supplementing, if necessary, with vitamins such as B12 or certain fatty acids.

Veganism is perceived by some as a huge fad, with bloggers becoming famous by sharing their vegan tips and recipes, and with restaurants and pop-ups thriving off the back of vegan diets. The information available about a vegan diet – or becoming pescatarian or reducing meat in general – is mind-boggling, but what you put into your body is ultimately up to you and how that food makes you feel. However, we should be grateful to the Internet and those willing to share their ideas. The wealth of vegan content online

▶

can help individuals decide what ideologies and values they choose to believe in. The vegan movement is not only attractive to those who want to learn about animal rights but is also awakening social consciousness about health and well-being and the environment.

Myth buster: being a vegan does not mean that your diet is totally sustainable

The reason veganism is not necessarily sustainable is because you could be eating soya beans from China or pineapples from Colombia, and those ingredients have huge carbon footprints. In the UK, vegetarian and vegan diets are becoming increasingly popular, so figuring out how to grow and meet demand for alternative protein sources is growing in importance. This means as a nation we need to start experimenting with new crops, for example, figuring out how to grow the likes of chickpeas, lentils and beans, which are typically grown on an agricultural scale for mass consumption, on a smaller scale. We also need to consider how they can be a match for our climate as a lot of these proteins can't cope with the frost common in the UK. Gaining knowledge and experimenting with alternative crops will be key for our growing population during times of climate change. The debate needs to be had that too fast a radical shift towards plant-based diets has its risks: not only could we fail to achieve realistic goals but also fail to recognise and embrace farming techniques that can give back to the land, encourage carbon capture, enhance biodiversity and support livelihoods.

The role of sustainable farming

At present, the vegan debate is fiercely strong with arguments on both sides dividing public opinion, but perhaps an even greater discussion should be focused around sustainable agriculture and organic farming. Sustainable agriculture is a method of farming that meets society's current food needs, without compromising the ability of future generations to meet their needs. In order to do this, growers may use methods to promote soil health, minimise water use, lower pollution levels on farms and use energy more efficiently. Sustainable agriculture also proves useful for others in the food supply chains, from retailers to consumers, who can support farmworker well-being, better food packaging and value environmentally friendly methods.

Organic farming can sometimes be classified as sustainable agriculture, but note that organic products can be cultivated on large industrial farms that are not certified.

Although the bottom line of Poore's study is that avoiding consumption of animal products delivers far better environmental benefits than trying to purchase sustainable meat and dairy, it also revealed a huge variability between different ways of producing the same food. For example, beef cattle raised on deforested land resulted in 12 times more greenhouse gases and used 50 times more land than those grazing on natural pasture.

The hunting and farming of meat has always been part of our human story, but the 20th century saw the application of the principles of the industrial revolution to agriculture, especially within livestock farming, and today if we suddenly transitioned to a vast need for plant sources of protein and specific vegetables we would need to industrialise there too. The Western calls to shift mainly to plant-based diets are misguided and could do more harm than good if we focus solely on livestock and don't consider the sustainability of all foods.

Rather than encouraging people to be seduced by exhortations to eat more products made from industrially grown soya, maize and grains, we should encourage sustainable forms of meat and dairy production based on traditional rotational systems and conservation grazing.

The reports by the Intergovernmental Panel on Climate Change on land use considered meat consumption as only one single element of an impartial look at global agriculture, and there were 'opportunities and benefits of resilient, sustainable and low greenhouse gas-emission animal-sourced food'.[5] I mentioned the ramifications of the industrial production of meat earlier, and there is no doubt that they are incredibly serious – but researchers and activists can make the mistake of assuming all meat, particularly red meat, is the same in terms of its impact, regardless of the way animals are farmed.

In the UK, permanent pastures account for 70 per cent of farmland, on which nothing else can really grow, and as such the land acts as a greenhouse gas sink. This means that grasslands across the UK absorb CO2 from the atmosphere and, as grass grows, carbon is sequestering in the soil as organic matter – and the more the land is grazed and trampled on by livestock the more it can absorb. This contradiction gives a unique perspective to the UK farming economy. Lots of media and documentaries are skewed towards US livestock practices, with UK methods of production being overlooked and its reputation tarred in tandem with the most extreme and intensive meat production systems in the world. While by no means perfect – there is industrialisation of meat production across the nation – British livestock is actually one of the most sustainable in the world. For example, 85 per cent of the water consumed by sheep and cattle falls as rain on pastures, and grass contributes significantly to the feed consumed by sheep and cattle – some reports suggest as much as 90 per cent.

Taking a regenerative approach

The regenerative farming approach focuses on restoring soils that have been degraded by the industrial agricultural system. Its methods promote healthier ecosystems by rebuilding organic soil matter through holistic farming and grazing techniques. There's a lot of ambiguity surrounding the exact definition of regenerative agriculture, but there is an opportunity for innovation around its core concept – promoting the circular economy and, in short, allowing nature to do the work. We should be questioning the ethics of driving up the demand for crops that require a high input of fertiliser, fungicides, pesticides and herbicides, while demonising sustainable forms of livestock farming that can restore soils, biodiversity and sequester carbon.[6] Through the use of rotations from green manure and animal manure, we can build soil quality, which, in the long run, produces heathier plants that are less susceptible to diseases and pests – crucial given the changing climate. Soils with high levels of natural organic matter improve water retention, which is vital in reducing flood risk and enhancing the ability of crops to withstand drought. Such all-encompassing farming techniques can also help restore biodiversity and allow wildlife to flourish again.[7]

The knock-on effect of large-scale veganism

Another factor to consider is that if the world went completely vegan, what would happen to millions of people employed in the livestock sector and what effects would a vegan 'epidemic' have on national economies? Unfortunately, there is no real way of monetising *not* eating animals. According to the Department for the Environment, Food and Rural Affairs, approximately 500,000 people in the UK work within farming enterprises with livestock, mostly in concentrated areas in the countryside. Over 17 million hectares of UK land are currently farmed, and it accommodates more than 10 million cows, 33 million sheep, 5 million pigs and almost 173 million poultry animals.

There is no avoiding the fact that, as a result of our omnivorous population, animals produce a considerable profit for the British economy. If we all went vegan, small-scale farms would be hit the hardest, and it wouldn't be as simple as everyone switching to growing vegetable crops, partly because the British countryside can't accommodate it. Over time we would find a way to replace the jobs for people formerly engaged in the livestock industry, who would need assistance transitioning to a new career, using subsidies, or alternative ways to use people's skill to avoid potential negative impacts on biodiversity. But we would need to consider very carefully what foods we could actually grow in the UK to feed people with. In the UK we don't grow enough legumes and nuts to sustain a fully British and balanced diet that is meat-free.

To farm arable crops, you need much more land of a particular quality. In the UK, many animals graze on land that isn't suitable for crops. Beyond the farms, without livestock entire rural communities would be pretty much wiped out. They need animals to keep the countryside going, but for now farmers might consider slowing down breeding as the demand for meat falls globally or adopting more sustainable ways of producing meat.

WHAT ABOUT CHICKEN FARMING?

The climate conversation is often dominated by the subject of cows, but in the UK we now consume more chicken than pork, lamb and beef combined, and last year over 1 billion chickens were reared. From major supermarkets to fried chicken buckets, the chicken sector is under pressure to satisfy the nation's growing demand for cheap protein.

As a result, large-scale commercial abattoirs and factory farms are working overdrive, and animal welfare is a secondary interest. Chicken is viewed as a commodity rather than a living species with needs, and chickens are reared indoors, their genetic traits modified so that by day 22 they resemble a fully-grown chicken and after just 44 days they are sent off to be slaughtered. These factories are capable of slaughtering over 30,000 chickens every hour.

For over 40 years, Daylesford Organic Farm in Gloucestershire has done things differently. One of the most sustainable farms in the UK, they are holding themselves accountable for the highest ethical and environmental standards.

At Daylesford, they don't agree with intensively reared chicken. As a result, they opened their own hatchery and chicken abattoir, certified to organic standards by the Soil Association. This means they control the whole process on their own farm. The organic parent flock lays eggs that are

▶

incubated in their own hatchery to provide organic chicks (at the moment there is only one other company in the UK that does this). As with their other animals, they choose slow-growing breeds with a natural instinct to explore and enjoy a varied foraged diet. After 70 days plus the birds travel less than a mile to their own abattoir, where as much of the process as possible is performed by hand.[8]

The demand for chicken is staggering and because of this, it is currently not possible to produce enough high-welfare, free-range birds to meet it. Fewer than 5 per cent of chickens in the UK are reared organically – which is why our choices matter so much and our intention to reduce our meat consumption is vital. If you decide to eat chicken, or any other meat for that matter, you as the consumer need to think about how that meat has been raised and choose accordingly.

But what about fish?

For over 3 billion people in parts of the developing world fish, not meat, is the only source of protein. The UN's Food and Agriculture Organization pointed out in a study just how reliant we have become on seafood as a source of protein. An estimated 10–12 per cent of the global population relies on fisheries and aquaculture for their livelihood. Per capita, fish consumption has risen from 10kg in the 1960s to over 19kg by 2012, and seafood production is increasing annually at a rate of 3.2 per cent – twice the world population growth rate.[9] Across the supply chain it is imperative that we begin to protect marine habitats, the species that live within them and the communities that depend on seafood caught and farmed with minimal environmental impact.

It's undoubtedly challenging for the world to become entirely vegan, and we don't yet know all the health implications, but attempts to overcome them are being made. Consumers like you can play a huge role in the positive reinforcement of a new food system by buying locally grown fruit and vegetables, and sustainably reared meat and fish from our waters, because this is the most balanced option. Climate change is deeply important, but so is the survival of every person on this Earth, so use your purchasing power as a way to support a better food system for you, the producers and the planet.

Myth buster: fish farming is environmentally friendly

We tend to have this obsession with beef and reducing beef consumption, but wild fish stock has decreased by 90 per cent in the last 50 years, and as a result fish aquaculture is helping to feed the increasing global population. Fish farming provides two-thirds of the fish sold in Asia and over 95 per cent in Europe.

Previously thought to be environmentally friendly, in fact, farmed fish are depositing excreta and unconsumed feed to the bottom of the ponds where there is barely any oxygen, making it the perfect environment for methane production and thus contributing to greenhouse gases.

Population growth and agriculture

At present, many of us have the privilege of choosing whether or not to go vegan, but there are some factors that are completely out of our control. Right now we live in a world where food is in abundance (although it isn't evenly distributed), but we are moving towards a very scary future where there will not be enough food or water for everyone. This will fuel economic crises, conflicts and water wars, because climate change poses a major challenge to the future of agriculture, which will need to cope with rising temperatures and a lack of water resources – all in the context of continued population growth.

By 2050 it's expected that another 2.2 billion people will join the planet. Feeding the world's population will be an enormous challenge, and as yet we don't know how we will achieve that. At present, at least one in nine people on Earth is undernourished or hungry because of our unequal food distribution (hunger is defined by the Food and Agriculture Organization of the United Nations as having not enough caloric intake to meet daily minimum energy requirements). To feed 10 billion people we need to increase global food production by at least 60 per cent. A great starting point would be to stop throwing away one-third of global food production and to consider new farming techniques that can adapt to changing climates.

Although it should be possible to feed the world, agriculture is, naturally, incredibly sensitive to climate fluctuations. Conditions for crop yields vary dramatically globally, and even more so with unusual weather patterns. A year of low rainfall, heatwaves or cold snaps at the wrong time, or extremes such as flooding, can have a significant effect on local crop yields and livestock production worldwide. But of course deforestation for crops and animal agriculture only compounds the problem of climate change.

Food insecurity around the world

If we don't start acting, we should be gloomy about hunger in the future, because our world population is rapidly growing. Agriculture is critical to fighting hunger, tackling malnutrition and boosting food security, but climate change is one of the biggest reasons for increasing global hunger rates.

Since 1970, hunger in the developing world has more than halved. The rate had gone from at least 35 per cent to less than 15 per cent, but in 2017 the trend reversed, and the number of undernourished people started to rise again. It is too short a time span to say whether this is a glitch or long-term trend, but it's worrying, and the UN says we won't definitively know whether the trends on hunger are temporary or not until after 2020.

Some researchers say climate change is exacerbating and even causing violent conflicts around the world. Almost a quarter of the populations of the least developed countries already suffer from food insecurity, with vulnerable populations in Nigeria, Somalia, South Sudan and Yemen facing the risk of famine in the coming decade. East Africa has seen years of drought, while South Asia has seen the worst flooding in years. Intensified fighting globally generally disrupts agriculture, making accessing food markets harder. It can even devalue a nation's currency, resulting in higher food prices, which places significant pressure on household incomes and the amount of money people have to feed their families.

Studies are also recognising that some sections of the developing world will never be able to sustain their populations with their own food production, and there is no doubt that climate change will cause further devastating food shortages at some point in the West as well.

Already, countries with huge populations, such as China, cannot physically grow enough food for their own needs, so they rely on food imports.

In America, farmers are making a living from genetically modified organisms (GMOs) on huge farms, covering hundreds of hectares of land, with the help of lots of expensive machinery and the use of agrochemicals. Across continents such as Africa, international companies are capitalising by removing traditional farming techniques in favour of GMOs, claiming that this will help to combat hunger. Millions of African farmers would strongly disagree, however, as they already have up-to-date approaches to seed agriculture, which are far friendlier than GMO. I won't get into how patronising Western ideals or corporations can be, but artificially scaling food production can result in a terrifying list of potential crises for the environment: climate disruption, mass extinction of wildlife, loss of biodiversity, pollution of our oceans and waterways, and the diminishing health of communities.

Growing practices that are suitable for the locality

Those who choose to listen will realise that African farmers have a vast amount of ecological knowledge. Their own varieties of seed have evolved alongside local pests and diseases and are adapted to different terrains and changing weather patterns. Communities in Africa that have experimented with GMOs using imported seeds, and with them the need to pump the ground full of toxic fertilisers, have experienced, and thus understood, that modern modified seeds seem to fail in times of changing climates and unpredictable weather. This highlights the misconception that industrial-scale agriculture is needed to feed a growing world population. These real-life situations, once again, debunk the widespread myth that we need chemicals, engineered seeds and factory farms to feed our growing population.

The only way to ensure real food security, both today and in the near future, for a growing population is to support farmers to revive their seed diversity, to maintain healthy, natural soil and to empower those who understand their own land. A global agricultural solution for all farming around the world does not exist, and each continent will have to adapt and trade. The large variability in the environmental impact from different farms does present an opportunity for reducing harm and feeding our ever-expanding global population.

THE QUINOA BOOM

The rise to fame of one of the world's oldest foods, quinoa, is an incredibly complicated story and one that highlights the complexity of meeting the ever-changing demand for crops. Quinoa is a seed called a 'pseudo-cereal grain' because it is not a grain, although it is used like one. Since health fanatics in the West recognised the nutritional benefits of this seed, quinoa has risen from being a local food to a global commodity.

The boom was an opportunity for farmers, in particular in Peru, and specifically the Andes. As demand for quinoa exploded, the price of quinoa boomed as well, making it an incredible opportunity for local growers to make significant profits from their crops and to rise out of poverty.

Before export demand increased, quinoa had been cultivated for over 7,000 years and was considered sacred by the Inca Empire. It's thought there were once 3,000 or so different varieties of quinoa, but, with increased demand, farmers began abandoning them to focus on only a few. This is problematic, because the varieties created by Andean farmers are the future of quinoa, and can adapt to the fluctuations of climate change. The soil quality in the areas where quinoa was grown commercially also began to deteriorate, because the high prices offered for the crop brought land into cultivation that previously had been allowed to rest, resulting in nutrient deficiency. Furthermore, farmers who began focusing on producing more quinoa started to reduce their llama herds, so there was less manure available as a natural fertiliser for the ground.

The high prices also meant that the poorer people in Peru, for whom quinoa was once a nourishing staple food, could no longer afford to buy it, and locals began switching their grain intake to cheaper imported junk food. There was then an incursion of large-scale farms outside the Andean highlands that began dashing the dreams of the local farmers by out-pricing their produce. Unsurprisingly, the high prices attracted competitors: in Peru, farmers on the coast began using intensive methods and fertilisers, getting double the yields of those in the Andes. Industrial farms flooded the market and, perhaps the greatest problem, consequentially the price of quinoa sank as fast as it had risen.

The story of quinoa's rise in popularity is an example of a damaging north–south exchange, where well-intentioned health ethics led to environmental degradation and further drove poverty in Peru. It doesn't mean we should stop eating quinoa, but it requires us to eat it or any other so-called superfood in moderation and not take it for granted.

Myth buster: the world can be fed only by large-scale agriculture

Most of the terrifying climate studies we read create a misconception that industrial-scale agriculture is needed to feed a growing world, but according to the UN, currently 70 per cent of the world's food is produced by small farm holders.

Agriculture creates jobs

Worldwide, the livelihood of over 2.5 billion people depends on agriculture, and it is central to economic development globally. Agriculture is known as one of the most powerful tools for raising poor people's incomes, but when most farmers are small-scale producers, they are, of course, the most vulnerable to environmental and price shocks. The unpredictability of crop yields because of climate change really does become a matter of life and death for these producers, and, in recent decades, they have had more than just a bad year or two. We need to start humanising our food chains and giving credit to the farmers and people who grow our food.

It is also worth noting that those in the farming industry depend on specific climate conditions. Right now, the same CO2 accumulating in our atmosphere in relation to fossil fuels is changing the composition of the fruits and vegetables we eat, making them less nutritious. The extra CO2 is rushing photosynthesis and causing plants to grow with more sugar and less calcium, protein, zinc and other important vitamins. The results in the future will not only be that people will become deficient in vital nutrients because of the changes to plant physiology, but there will also be changes in the ozone layer and greenhouse gases, and climate change will affect the livelihoods of farmers due to the unpredictable nature of crop cycles.

In developing parts of the world, where people depend on a decent harvest to feed their families, earn an income and pay back debts, extreme heat and other disruptive events are starting to trigger an existential crisis. In some cases, the emotional toll can become too much to bear, and climate change is thought to be one of the primary triggers in a rise in farmer suicides since the 1980s. Obviously the agricultural industry is incredibly vulnerable and climate change is not the only factor contributing to declining mental health and higher suicide rates amongst farmers, but they are on the front line and are most vulnerable to changes in temperatures and rainfall, as well as frequent extreme weather events. Time to Change Wales has learnt that at least one farmer with a mental health problem takes their own life each week in the UK,[10] meaning more than 52 families every year will carry that tragic burden with them for the rest of their lives. Further afield, the farmer suicide epidemic in India is a particularly distressing problem right now, as temperature spikes and frequent flooding have resulted in crop losses.

The story of Vishal Pawar

This is an incredibly sad story, but it is one that is becoming more prevalent as small farmers struggle with climate disasters.

In 2010, Vishal Pawar left his job and established himself as a farmer on his own 15 acres of land. His previous work experience and ideas about modern farming methods helped him to build a well to provide water for his land. In 2013, Vishal's well dried up due to severe drought in the region. India had been experiencing a series of traumatic weather events, including extreme heat and intense flooding across the country. Vishal attempted to fix his well, but with no luck; he was left with no other option than to take an extra 10-acre area of land on lease and start farming that instead. His plan was to work hard on the extra land and repay all his loans as soon as possible.

That same year a huge monsoon hit his region. Climatic conditions did not favour him, and his land reaped only 10 per cent of what he expected. Feeling trapped due to economic despair and with no food to feed his family, Vishal sadly took his own life. The plea below is part of the note he left:

To,

Respected Guardian Minister.

Sanjay Rathod Saheb, I have not been able to reap my crop due to bad Monsoon.

There is Drought. I was therefore not in a position to repay the Bank and my Debtors. I am totally bankrupt and broken.

I am unable to take the responsibility and look after my family. Please do ensure to take off the debt when I am no longer there. This is a plea . . .

Vishal Pawar

Appreciating the important work of farmers

Vishal's story is just one of thousands of farmer suicide stories that have been reported in recent years.[11] A suicide epidemic among India's farmers, and in other countries, has shaken the industry severely. It is a widespread and intensely personal issue, so it is difficult to tease out the root cause, but a combination of debt, lack of social services and support, and recent climate change have intensified the situation for small-scale farmers.

We must support small-scale farmers like Vishal, by ensuring suitable crop insurance and compensation of losses due to climate-related factors, thus avoiding the sense of hopelessness that leads to despair and even suicide. Farmers need resilience, smart strategies and specific skills to live well and farm well. Countries must start using data to continuously improve the efficiency of their agricultural supply chains, enabling different sections of the supply chain to monitor environmental and other conditions.

The data can be used to adapt how crops are produced, stored and distributed to reduce waste. We must support science and technologies that can generate new crops that can adapt to new conditions. Food producers will benefit from data-driven farming, such as checking satellite images and sensors, gathering vital information about weather patterns, soil conditions and market demand with just a few taps on a mobile phone. Climate policy for the future will also be essential to combat extreme hardship for farmers, not only in the Southern Hemisphere but worldwide.

Looking ahead at future generations worldwide, low earnings, high land prices and a shortage of rentable farmland are discouraging young people from joining the farming industry.

Without a doubt, farmers have felt beleaguered as our world shifts perceptibly towards dairy free, vegetarian and vegan lifestyles, and they are faced with what can feel like a prolonged criticism from the media on the impact of farming on climate change. Caught in the middle, farmers also have to face the fact that the majority of British households are, for now, more concerned with price than issues of sustainability. In such a situation the danger is that the best quality British produce becomes a staple of the well off, while lower income households make do with cheap global imports that don't meet safety or environmental standards.

Many farmers are starting to supplement their incomes by diversifying activities away from pure agriculture. Biofuels, for example, present new opportunities against a background of rising fears about energy security, rising fuel prices and climate change. But there must be more awareness, and that includes at home in the UK, that farmers have an important role to play as custodians of the countryside and biodiversity. We need future generations to be involved, or more engaged, in the agricultural sector. To keep future farmers safe, we must be mindful of where and how our food is produced, give thanks to our farmers, support local produce and ensure every bite counts, no matter where we are in the world.

Tying this back to farmers in the UK, it's not fair to shun sustainably produced British meat in favour of 'plant-based diets' of avocados or almonds that are grown in some of the driest places on Earth using precious water sources. Can British farmers do more? Yes, of course! To reach the point where British produce is the first food choice for all in the UK we need farmers, consumers and governments to make climate-friendly food a primary concern. Our environment and our health depends on it.

Food waste: why are we so inefficient?

There is nothing worse than opening the fridge door, reaching in and pulling out the remains of a dinner that's now gone off. The disappointment is major, not to mention the disgust when throwing away the rotten items.

What do you do when the salad bag in your fridge still looks fine although its use-by date has passed? There are absolutely no visual signs or even nasty smells to suggest the salad is bad. If you're like most people, the better-safe-than-sorry instinct kicks in and you throw it away, telling yourself that you will be more mindful of dates next time. In the UK more than half the food we throw away is food we could have eaten. As individuals at home, we play one of the biggest roles in contributing to this issue. About one-third of waste occurs at the consumer level, where we buy too much for our needs and throw it away. Our emotions and brains are conditioned to live by the end date stamped on produce. An estimated 25,000 tonnes of the food that goes to waste each year is still edible; that is approximately 650 million dinners that could have been made from perfectly edible leftovers.

Food is plentiful but unevenly distributed

Around the world, over 800 million people will go to bed tonight chronically under-fed; 200 million will be children, and they will live mainly in rural areas. Shockingly, though, there is currently enough food on the planet to feed everyone. We could technically feed all 800-plus million people on less than one quarter of the food that is wasted in the US, the UK and Europe every day. As of now, food supply is not the issue; it's the distribution. Unlike water or oil, which are finite resources, we grow enough grains, such as wheat and rice, to give every person on the planet 3,500 calories a day – about 50 per cent more than an average adult requires in a day.[12] What happens instead is that we poorly distribute and waste ridiculous amounts of this food.

Food distribution is currently unequal and we must start thinking ahead. Ending hunger in a sustainable way is morally right, politically beneficial and economically feasible for those in positions of power.

The problem of food waste

Food waste is a humanitarian issue as much as an environmental one – we must focus on how to feed more people while wasting less of what we already produce.

Food waste levels vary in every part of the world but if global food waste were a country, it would come in third after China and the US in terms of its impact on global warming. These numbers are predicted to rise by one-third by 2030, when 2.1

billion tonnes will be either lost or thrown away, equivalent to 66 tonnes per second.[13] At the production and distribution level, for example, food deteriorates in fields, or is lost as a result of poor transportation, or it spoils in markets that lack proper preservation techniques.

Understanding where waste occurs along the supply chain provides an opportunity for effective solutions to be implemented. The sheer amount of food waste also has a significant impact on climate radiations because of the levels of carbon dioxide, methane and nitrous oxide that are produced and emitted throughout the various stages of food production and onwards to when we put it into our refrigerators. The energy that goes into the production, harvesting, transportation and packaging of any food wasted generates more than 3.3 billion tonnes of carbon dioxide. In addition, there are all the emissions created from producing the metal cans, plastic bags and cardboard containers that our food comes wrapped in, as well as storage, transportation and cooking that are also wasted when food isn't eaten.[14]

Supermarkets and food waste

There is a lot going on throughout food supply chains that is mostly out of our personal control, but what about tackling the supermarkets, where most of us buy our food? In monetary terms, the world loses $400 billion of food before it reaches the supermarkets. An estimated 20–40 per cent of UK fruit and vegetables are rejected before they even reach the shops. Do the top large supermarkets really need to order so much food that they won't sell? Do supermarkets really need to provide us with hundreds, if not thousands, of differently branded but similar items? Is it necessary that we have five instant rice brands on one shelf, or ten brands of the same tomato sauce in different packaging? As consumers, do you feel that you need rows and rows of different brands?

In addition to supermarket supply mismanagement, what about the misuse of labels and best-before dates imposed on food, making consumers paranoid? Supermarkets need to improve their labelling standards or even remove labels on some produce, allowing the consumer to better understand the freshness of products and avoid unnecessary waste.

Food waste is a relatively easy problem to solve if we can become more efficient and smarter about the way producers distribute and what we consume. As consumers our responsibility lies in buying only what we intend to eat. Accept the imperfect, embrace the ugly and use your intuition as a consumer of what's good or not to eat.

The climate is changing – you can too

On a personal level every bite counts, being conscious of your food's carbon footprint can greatly help the rest of the world. The decisions we take every day on the food we purchase is the single most important contribution you, as an individual, can make.

Do your research and ask yourself the simple question: where does this food I am buying come from?

Accept that produce can be of perfect quality, delicious and safe, even if it has a slight imperfection in appearance.

Support local farmers Support organically grown produce or regenerative crops.

Buy local Visit a farmers' market. You will most likely be surprised at how much more flavoursome local produce tastes in comparison to international/supermarket-bought veggies. If you don't have time to visit any local farmers' markets, sign up to delivery services such as:

- Oddbox. The social enterprise, Oddbox, is on a mission to slash the amount of fruit and veg that goes to waste and landfill because of minor imperfections. Each box comes with a guide outlining what food has been rescued from both UK farmers and abroad.
- Abel and Cole. Deliveries contain all organic produce from British farms, that changes with the seasons.

Buy organic Organic means farming without chemical fertilisers, most pesticides, genetic modification or the routine use of drugs or antibiotics. Some of the pesticides on non-organic fruit and veggies can remain even after they are washed properly. In a perfect world, buying everything organic would be ideal, but convenience and cost may not permit that. According to the Environmental Working Group, here is the 'clean' safest non-organic produce to buy, and the 'danger' fruits and veggies that you should buy as organic wherever possible, because they normally contain the highest concentrations of pesticides.

- Clean: asparagus, aubergine, avocado, broccoli, cabbage, cauliflower, kiwi fruit, melon, mushroom, onion, papaya, pineapple, sweetcorn.
- Danger: apple, celery, cherry, grape, kale, nectarine, peach, pear, potato, spinach, strawberry, tomato.

There is absolutely no question that we should all be eating far less meat, and calling for an end to high-carbon, polluting, unethical and intensive forms of grain-fed meat production.

Adopt meatless Monday Founded by Paul, Stella and Mary McCartney, Meat-Free Monday is an environmental campaign to raise awareness of the climate-changing impact of meat production and consumption. The message is simple: one day a week, cut the meat.

If you do go vegan, it doesn't necessarily mean you're eating healthily, though. There are a lot of processed vegan junk-food equivalents – so make sure your diet is rich in all the essential nutrients. It is important to include proteins from foods such as nuts, seeds, beans and lentils.

Bread is one of the most unnecessarily disposed of food items Don't contribute to the 24 million slices of bread wasted every day in the UK: freeze your loaf and defrost it a slice at a time, then whizz it into breadcrumbs, make croutons or turn it into delicious puddings.

If we look at the carbon footprint of various foods, as in how much do different foods contribute to the greenhouse effect, some of the worst offenders include: beef, lamb, cheese, pork, farmed salmon, turkey, chicken, canned tuna and eggs. Foods that are better for the environment include broccoli, citrus fruits, dry beans, green peas, lentils, nuts, rice, tofu and tomatoes. To reduce your food carbon footprint further, remember to source as much of your food locally as you can.

When eating fish or seafood look out for Sustainable Certified Seafood badges both at the supermarket and at restaurants. This ensures the seafood is either caught or farmed in ways that consider the long-term viability of harvested species and the well-being of the oceans, as well as the livelihoods of the fishery's dependent communities.

Be creative with your leftovers. There are tons of creative online chefs and bloggers encouraging you to minimise your food waste. Could you use your freezer more? Frozen foods can be just as nutritious, and they stay edible for longer.

At the supermarket avoid foods packaged in polystyrene and PVC, because it is unlikely that any of these materials will be accepted and recycled by your local council.

Sign up to collection schemes to give away your leftovers to those in need. Start using apps like Olio. Olio connects neighbours with each other, and volunteers with businesses, so that surplus food and other items can be shared, not thrown away. Olio is for food lovers who hate waste, care about the environment and want to connect with their community.

Spread the word about diet switches and veganism to your own circle of friends and acquaintances.

Watch some insightful documentaries about our food system such as:

- *The Game Changers*
- *Food, Inc*
- *Cowspiracy: The Sustainability Secret*
- *In Defence of Food*
- *What the Health*
- *Sushi: The Global Catch*
- *More than Honey*
- *Vegucated*

5

It's a Lifestyle, Not Just a Movement

'Nature is not a place to visit. It is home'

Gary Snyder

Everyone faces choices every day that carry a climate cost: from what we feast on in the morning to how much we keep our lights on at home.

Between 60 and 80 per cent of our impacts on the planet come from our household consumption. This includes many overlapping factors: from the food we cook to the clothes we wear and wash, and the energy we use to power our homes and appliances. The person who walks the climate walk is most likely not perfect, because they might be a vegan who still flies or a mum who has no toxic cleaning products at home but doesn't use reusable nappies. The level of sacrifice necessary to live a totally sustainable life is unacceptable to most, and we don't have the luxury of time or financial means to always opt for lower carbon options. Although we need systematic changes for the environmental battle, personally living sustainably is a journey.

Making significant lifestyle changes won't happen overnight either, and so, as you begin your process, this chapter aims to debunk and shatter some of the difficult misconceptions about living a more waste-free lifestyle at home and in your everyday actions. It will touch upon babies, energy, miscellaneous house chores, cleaning methods, composting and air travel. The best advantages of an eco-friendlier lifestyle are that it makes life simple again, it is cost effective, timely, fun – and entirely possible for both personal and family life.

Only you can drive the change by curating your home and life in the way that you can, to make sustainability a lifestyle, not just a movement.

The hard facts: home and lifestyle

- In the UK 90 per cent of people claim to recycle regularly from the kitchen, but only 52 per cent say they do so from the bathroom.
- Heating accounts for more than 67 per cent of household energy consumption in the EU.
- Eighty-four per cent of the public have demonstrated support towards renewable energy.
- One-third of UK electricity came from renewable sources in 2018.

- Over 25 per cent of our daily waste is compostable.
- A short-haul flight from London to Edinburgh contributes more CO2 than the mean annual emissions of a person in Uganda or Somalia.
- The average household does almost 400 laundry loads each year, consuming about 13,500 gallons of water. Ninety per cent of the energy used to power each wash load is used to heat the water. Only 10 per cent of the total power is actually used to power the motor.
- In the US over 2 billion plastic razors are thrown away each year.
- A baby uses an estimated 4,000–6,000 nappies before he or she is potty trained and these will cost families close to £750–£1,000.

The home

A green home could be a house designed to be environmentally sustainable, focusing on the efficient use of energy, water and building materials. Home is also where you spend most of your time so it is where you can begin to contribute to a sustainable life, however small. From the cleaning materials you use, to laundry, composting and decorating your home, being eco-wise can go a long way in saving you money and helping the planet, too.

The impact of cleaning products

Being aware of the ingredients in your personal hygiene products and your cleaning products is essential for the safety of yourself and your home. Cleaning up our planet could start simply with cleaning our homes. When you put your food down on your kitchen counter top, and when you put your toothbrush down in the bathroom, they are both most likely going to pick up the remains of the cleaning products that you last used. Have you ever taken a moment to read the ingredients of what is in bleach, surface sprays or glass cleaners? Most of them contain highly toxic chemicals that can cause skin irritation, organ toxicity and a bunch of other problems. What's more, these conventional products are sold by the hundreds of billions in single-use plastic bottles.

DIY CLEANER

Home is supposed to be a safe place for you and your family, so if you don't like the thought of your safe space being riddled with severe chemicals, you can make your own truly *all*-purpose non-toxic cleaner. The recipe could not be any easier to follow.[1]

You will need:

- Glass spray bottle
- Water

▶

Distilled white vinegar or apple cider vinegar
Scented essential oil (optional)

Fill a glass spray bottle half-full with water. Fill the remainder with distilled white vinegar. Add 15–20 drops of essential oil for scent. Spray away.

Apple cider vinegar or distilled white vinegar, bicarbonate of soda and citrus fruits all work great to clean areas at home.

Purchased environment-friendly cleaners

If making your own sounds like a lot of work, then support and shop from a company such as Method UK, the world's largest green plant-based cleaning company. Easy to buy online, their biodegradable plant-based ingredients are tough on stains, are contained in 100 per cent recycled bottles, and are only tested by people and never on animals: www.methodproducts.co.uk.

Start-up firm Blue Land are revolutionising cleaning to eliminate single-use plastics and toxic chemicals. Albeit only in the US for now, but they are rapidly expanding. The way their refill system works is: you buy your one off 'forever bottle', a non-leaching, BPA-free and shatterproof bottle. Fill the bottle with tap water, then drop in your cleaning tablet and clean.

When all your cleaning mixture finishes, you just refill with water and another cleaning tablet, and go again. Blue Land are putting the circular into cleaning: www.blueland.com.

You might also consider questioning what you use to wipe down those pesky surfaces with your alternative cleaning products. Ghastly amounts of kitchen roll or tissues? Or those antiseptic wipes? Absolutely not! They are the worst things you could use, because they just create unnecessary waste that ends up in landfill. The supermarket microfibre towels are full of plastic and non-recyclable, and if you wash them, of course, they are releasing microplastics into our water supplies, so, what's the best alternative?

- Buy cotton rags if you can.
- Tear old cotton clothing that is not suitable for donation, and make it into a cleaning cloth.
- Do you have old underwear with nowhere to dispose of it? Use it to clean.

The eco guide to laundry

The average household does almost 400 laundry loads each year, consuming about 13,500 gallons of water. Around 90 per cent of the total energy used by a typical washing machine is used merely to heat the water, only 10 per cent is used to power the motor. And, apparently, only 40 per cent of the clothes we put in our washing machines are dirty. How, then, can we stop being wasteful in the laundry department?

We humans have a tendency to throw an item of clothing in the wash because we've worn it once, because we feel that perhaps we should, just in case, because that's what we have been taught to do most of our lives. Many of us also have the awful habit of throwing a couple of items into the machine for a full wash. Studies have shown that you can save upwards of 45kg of carbon dioxide per household every year by running only full loads of laundry.

Hot washes and drying cycles also use huge amounts of water and energy to operate. By switching to eco-qualified and energy-saving approved machines you could save as much as 7,000 gallons of water per year. Over the approximately 11-year life of a washer, that is the equivalent of a lifetime's drinking water for six people.[2] Most washing machines have eco-friendly programmes, so try to use this option when washing clothes – even though the wash will take longer.

When it comes to energy, turn your machine down to 30°C. By cutting out the dryer, you will also save considerable sums on your utility bills. Buying a drying rack or hanging your items outside to dry might sound old-school, but you can save up to 320kg of carbon dioxide, and it also means less wear and tear on your garments.

Hundreds of thousands of plastic microfibres are shed every time we wash clothes, as I explained in Chapter 3, and these contain materials such as nylon, polyester and acrylic. They end up in the water supplies and ultimately drain into our oceans and are consumed by marine life. To put plastic fibres into perspective, a city the size of Berlin might be responsible for releasing the equivalent of 540,000 plastic bags-worth of them into the ocean daily.[3] That is utterly crazy, but while microfibre filters for washing machines are still in development, your most viable solution, other than using only non-synthetic textiles, is to buy yourself a mesh laundry bag, aka a Guppy Bag (as described on page 86).

Lastly, commercial laundry detergents and chlorine bleach contain a load of bad chemicals that can be extremely toxic to your skin and affect your respiratory system. You can make your own or try to buy an eco-friendly brand wherever possible. Here's a summary and a few other tips:

- Wear your clothes more than once.
- Use greener and safer laundry detergents. Smol products, for example, have some of the most effective concentrated laundry capsules on the planet. All are vegan and small enough to be posted through your letterbox upon delivery. They also work as a subscription service and can help you to save up to 50 per cent on your normal brand price: www.smolproducts.com.
- Turn the washing machine down to 30°C or below.
- In exchange for fabric softener, throw in a cup of vinegar. The pH of vinegar helps to neutralise the pH of your wash, leaving you with the fluffy goodness of your clothes.
- Wash delicate garments by hand.
- If you live in a big cosmopolitan city, head to the launderette. Commercial washers and dryers tend to be more efficient than domestic versions.
- Cut the dryer out whenever possible and air-dry garments.

- Save even more energy by not ironing wherever possible. (Honestly, since college and pre-getting into environmentalism, not ironing has mostly been a policy in my own life.)
- Try new sprays such as Day 2, the world's first ever 'dry shampoo for your clothes' so that you can freshen up your garments just like your morning hair. Day 2 prevents you over-washing your garments and could in turn save thousands of litres of water a year.

Composting

'Information is like compost; it does no good unless you spread it around'

Eliot Coleman

Compost is essentially nature's way of recycling by biodegrading organic waste. It is the breaking down of organic waste into soil over time, using heat and worms. Compost is broken down by the bacteria in the mixture, the creatures, such as worms and slugs, as well as the heat. The final product is a dark, crumbly substance with an earthy smell.

In Chapter 4 I spoke about carbon emissions connected to food production. As we have seen, throwing away food not only wastes money, but it also harms the planet far more than most of us realise. As individuals at home we play one of the biggest roles in contributing to this issue. In the UK, over two-thirds of adults have no idea that food waste contributes to climate change, while an even larger percentage (81 per cent) do not know that a third of all food is wasted globally.[4] Most of us who cook always have a load of rubbish after preparing meals, or we throw out uneaten food because of expiry dates. Even though natural food waste, such as fruit and vegetables, is biodegradable, it unfortunately won't biodegrade or break down as it is supposed to in a normal landfill pile – it needs specific conditions with lots of oxygen and heat.

Composting seems complicated, but it isn't and it is a great habit to get into because it keeps excessive food out of landfill, which in turn reduces methane gases (remember, methane is many times worse than carbon dioxide). Compost is mainly used to help keep our soil and plants healthy and it reduces the global need for chemical fertilisers.

OK, so it may sound as if you must live on a farm in the middle of the countryside to be able to compost, with lots of land and animals to make those oxygen and chemical reactions happen. There is a much simpler way, though, and it is completely possible, no matter where you live, even if it's a tiny apartment in the middle of a city.

Composting at home or away

Firstly, you need to decide whether you will be able to use your own composter or someone else's, such as the local authority:

Option 1: get someone else to do it for you There are a bunch of organisations and companies that will take your compost away and compost it for you. This is for everyone living in a city or urban populated place with not much outside space to break down the waste yourself.

Check in with your local council – they may do it for you. If not, check out your local farmers' market, as most likely there will be someone there who will take your piles of food waste off your hands. If neither of the above are viable, just Google it. Some companies will collect, whereas others might require you to drop it somewhere nearby. Each company will most likely vary in what they will and won't accept for composting, so read the guidelines.

Compost storage

How will you store your waste before collection day? If using a composting system outside the home, store your raw waste (as listed below) either in an airtight container or in a bowl or paper bag in the freezer to collect the waste. The freezer stops the smell of waste and anything leaking – it is arguably the best way to collect all your compostable waste. If you have your own composter, use a lidded container that you can empty every day into the composter. Whenever you cook or eat fruit, just get used to tossing your waste into your compost container or frozen bowl.

Option 2: garden composting This is something I have never practised from my little London apartment, so, if you are truly interested, I encourage you get lost in the online plethora of compostable content. If you just want the basics, though, here is the easiest way. Buy a compost bin (preferably two). Although they are plastic, they are better than making a pile in the garden because this will attract rats or other wild animals due to the quantity of food scraps in the compost. Put it in your garden, then put all your uncooked waste in the bin, plus your garden waste (cuttings, weeds and so on) and leave it to break down – it will take a much shorter period of time in the summer because of the heat. If you have two bins, you can use the other bin while the first one is composting. The earth attached to garden weeds adds to the mixture and introduces microbes, insects and bacteria to aid the process. You can look up on the Internet how to make perfect compost, but otherwise you will probably have compost that is not bad at all just by following this simple routine. Never put cooked food into the compost bin.

What can you compost?

Now you need to know what counts as compostable and what doesn't. This is not a list of absolutely everything, and each compost pile will vary in ingredients. Remember: only ever put raw materials into the compost:

- All fruit and vegetable food scraps
- Seeds
- Grains
- Egg shells

- Coffee grounds and loose tea leaves
- Old wine (it's made of grapes, remember)
- Paper
- Leaves and garden waste such as grass and weed trimmings

What about the bathroom?

In the UK, 90 per cent of people claim to regularly recycle from the kitchen, but only 52 per cent say they do so from the bathroom, according to the Recycle Now organisation. Have you ever thrown away a loo roll, a perfume bottle, foundation tube or plastic toothbrush without thinking? Guilty. It can be so confusing, because not even the brands selling the products are advising us what to do on the packaging, and many don't contain the green recycle logo either. Some of the most common bathroom offenders are noted below and they are easy to switch out.

Save the trees and buy paper made without them

Your average toilet paper is often bleached and comes in plastic packaging, making it high in toxic chemicals overall and non-recyclable. Most toilet paper is still made with virgin trees. Not only are we cutting them down but we are also using an enormous amount of energy and water along the way to clean our body parts and home surfaces. Here are alternative ideas:

- Bamboo paper derived from bamboo forests is a great sustainable alternative to paper. Bamboo growth uses less water, generates more oxygen and removes more carbon dioxide from the air than trees, making it the most renewable source of plant-based material available for manufacturing products on Earth.
- Shop online from brands like Who Gives A Crap. Their delivery service is extremely convenient. They started their business after learning that 2.3 billion people across the world don't have access to a toilet. That's roughly 40 per cent of the global population, and it means that around 289,000 children under five die every year from diseases caused by poor water and sanitation. Today, not only their toilet paper and paper rolls grace our bathrooms without harming any trees in the process, because they are made from bamboo, but the company donates 50 per cent of their profits to help build toilets and improve sanitation in the developing world.[5] Uk.whogivesacrap.org

Plastic toothbrushes

Over one billion plastic toothbrushes are thrown away every year in the USA alone. Be future-friendly by switching to bamboo toothbrushes, or an electric one. Bamboo or wooden toothbrushes come from sustainable sources, but please read the small print, as the bristles can often be plastic, and check that the bristles are truly compostable.

Toothpaste

There are many innovations happening on how to make your toothpaste plastic-free, by offering it in paste form in glass jars or as tablets. For example, there is an all-natural way to replace the paste you've used your whole life. 'Bite' is made with vegan-friendly ingredients like coconut oil and xylitol with natural whitening properties. Just pop one 'bite' in your mouth every time you brush your teeth. They come in refillable and recyclable glass jars.

Razors

In the UK, in 2018, an estimated 5.5 million people used, on average, multiple disposable razors. A simple swap would be to use metal safety razors instead of plastic ones.

Soaps

Switch out all liquid soaps, shampoos, and so on, and either buy in bulk to transfer to your own containers or find hard-bar versions to dramatically save on plastic pollution.

Recycle

Wash all glass containers before disposing of them. Most spray cans once completely empty (apart from hairspray) are recyclable. Just remove any lids and put them into your regular recycling bag for cans.

Smart energy

A future where almost all our electricity comes from wind, wave and sun is not actually an unimaginable scenario. In the UK in the past six years we have upped renewable electricity output from 7 per cent to 25 per cent. Technology is developing rapidly across the world, allowing costs to fall at a lightning pace, and there are billions of dollars invested in renewable power. Since 2009, there has been a 90 per cent drop in solar costs globally. As a result, there is an appetite for more renewables demonstrated by the public, and 84 per cent of households in the UK said that they would be happy to support renewable energy.

ELECTRICITY SURGES

Our national grid is used to coping with surges and falls in demand for electricity. Over the decades, as population increases and more houses are built, you might assume that would put more pressure on energy supplies. In fact, at half-time in England's 1990 World Cup semi-final, millions of fans brewed half-time cuppas. This caused the largest spike in demand for electricity to date, at a power-popping 2,800 megawatts.[6]

Our grids are becoming smarter in order to match supply and demand, and by making green switches in your home you will also save yourself lots of money. In other parts of the world, from South Africa to California and Chile, wind and solar are now the cheapest sources of new electricity, and, don't worry, our weather monitors are so smart that when there is no sun shining or wind blowing, you won't lose power at home. In the UK, the 2020 renewable target of 15 per cent set by the government has been met with widespread criticism, with many accusing the government of not doing enough. But are we right to criticise when we don't even know what is possible in the future? A study showed that it is possible for the UK's power system to be nearly 90 per cent renewable by 2030, while electrifying 25 per cent of all heating demand and putting over a million electric cars on the road, but only if we can cut demand for heating by almost 60 per cent over the next 15 years – that is a major challenge.[7] Reaching such high targets in the UK, and worldwide will require shared responsibility and huge investment.

Every home owner or renter across the UK and in different parts of the world, even those living in really old buildings, have many techniques at their disposal to turn their properties into greener establishments.

Upgrade your lighting

Incandescent bulbs were phased out in the UK in 2009, thanks to European regulations, as they were branded environmentally-unfriendly. While many people preferred the warmer glow given by incandescent bulbs, 95 per cent of the energy that goes into them gets turned into heat rather than light. They were replaced by LED and fluorescent bulbs that are far more energy efficient. Not only do they require much less electricity, but most have longer life cycles. LED light bulbs also don't use mercury. After all, a single low-energy light bulb, available for around £1, can, over the space of six years, save 250kg of CO2 – equivalent to that of a short flight.[8]

Invest in energy-efficient appliances

When it's time to renew big appliances such as refrigerators, washing machines, dryers and televisions, choose more energy-efficient models. There are many available now that have been certified as energy-efficient, which is normally confirmed by an energy star on the packaging label, so opt for those whenever possible. High-energy appliances increase the amount of carbon emissions and other waste resulting from the operation of power plants, which ultimately lead to increased amounts of toxic waste, air pollution and water pollution.[9]

Green electricity

In literally five minutes on platforms like uSwitch you can switch to a green energy provider without any disruption to your supply. By choosing a green tariff you are showing your support and joining the movement for more renewables.

Home décor

Now we move away from the miscellaneous chores and on to lifestyle. Whether you are hosting a birthday celebration, a dinner party with friends or decorating for Christmas, the best way to save on cleaning and to save money is to keep it simple. Keeping it simple in no way means boring – it means doing away with the tacky £1 decorations, trinkets and glitter-dropping baubles. For goodness sake, you don't need to be using valuable space hoarding a bunch of decorations that get used once or twice a year. Instead, turn to Mother Nature and use more natural and organic materials that will give your home a chicer look.

- Use natural materials for décor items. Think natural, such as branches, flowers, hemp, and 100 per cent organic. Fill glass jars with cinnamon sticks, pinecones and dried leaves.
- Search for sustainably sourced wax candles to set the mood for any event.
- If you don't already have a plastic Christmas tree that you use and can continue using, buy a real tree this year. If you don't have anywhere for the tree post-use to be replanted, most communities offer recycling resources where trees will be used for community projects. Whatever you do, try to avoid it being left for the rubbish man to pick up and take to landfill. Some local councils have a special tree-collection service. Christmas trees are beautiful renewable resources.
- In case you don't want to go the natural way and want to

decorate your Christmas tree or other areas of your house, find second-hand decorations to buy either online or in vintage stores.

- Use fewer twinkly lights unless powered by solar or another renewable form. Lights can be beautiful but they suck up a load of energy, so unless you really need them, at least opt for fewer – your electricity bill will also thank you.
- The same principles apply to fancy dress parties such as Hallowe'en. Most costumes contain so much plastic, are expensive and are worn only a handful of times. Try combining existing clothing items with pieces from charity shops and swapping with friends, or make dressing-up a fun activity with kids and make your own costumes at home with existing household goods.

Nature baby

The biggest decision many of us will ever face is whether to become a parent. The drive to reproduce is ancient, but a startling controversial view is that having children is technically the most destructive thing a person can do to the environment. In the era of climate change the number of families considering the moral decision not to have children is a conversation of its own. Around the world there are an average 360,000 births per day – that's 250 babies each minute.

The impact a child has on the environment also depends on where you live. The carbon impact of a child born in the

US, for example, is more than 160 times that of a child born in Bangladesh. In Europe, children have emissions not far from the global average, and in the UK the average per person is around 5.8 tonnes. If you are expecting a baby, or you already have kids, especially toddlers, adding sustainability to your life needs to be simple and sane when your little ones already require your undivided attention.

Although the climate news might not compel you to adjust your family size, hopefully some of the information in this book might motivate you to adopt some family lifestyle changes to reduce your environmental impact.

Being a climate-conscious parent can include sourcing quality organic products, finding new uses for things that you own, or passing on items to other children or to charity, thus keeping your child's carbon emissions low. And raising educated and motivated children who are compelled to fight for political progress on climate change and human rights is important too.

Nappies

A baby uses an estimated 4,000–6,000 nappies before he or she is potty trained and it will cost families close to £750–£1,000 for these non-recyclable bum-huggers, which take 500-plus years to decompose in landfill while creating a lot of CO2 emissions. In comparison, babies need on average between 20–25 reusable nappies. Reusable nappies work out cheaper, can be saved for future kids or donated to friends or charity, so the environmental impact is much lower.

A reusable nappy does sound a little scary, but a lot of people tend to make the switch progressively, starting when it's convenient. The nappies need to be washed every two to three days with hot water, removing the solids first into the toilet.

The most sustainable type of nappy is a recent innovation by Luisa Kahlfeldt, known as a sumo nappy and made entirely from seaweed and eucalyptus fabric, called SeaCell. The materials are sustainable to harvest and it is really easy to recycle after use, with no need to disassemble its components. The textile is antibacterial and antioxidant-rich, so it's beneficial for a baby's skin. If a sustainable nappy is just not for you, look out for disposable brands that are starting to commit to reducing their environmental impact, such as sustainably sourced materials or reduced water and energy emissions during the fabrication.

Baby wipes

Although they might offer momentary convenience, baby wipes cause havoc. They take hundreds of years to decompose and often end up in sewers, and then our oceans. Ditch disposables and use cloth wipes or reusable wipes, or cut up old muslin cloths and repurpose them.

Clothes

Babies grow incredibly fast and can't actually voice an opinion about what they wear. If buying new clothes, try to opt for garments made from natural fibres by ethical brands, or use pre-worn clothes. If you already have a ton of baby clothes at home that you no longer use, donate them to family, friends or a charity.

Breastfeeding

Breastfeeding is remarkably green and, in fact, it has been called the most environmentally friendly food available on Earth. It produces zero waste, zero greenhouse gases and has zero water footprint. If breastfeeding is an option for your family, you will be saving considerable amounts of water and energy that formula milk requires.

Toys

Value experiences over 'stuff' with your kids.

Travel: flight shaming

From eating less meat to using less plastic, we have seen the many ways that we can reduce our carbon footprint – the total amount of greenhouse gas emissions that we have produced through our actions. The greenhouse gases are the gases in the atmosphere that contribute to global warming and climate change, and although we know that almost every aspect of our life has a carbon footprint, a huge chunk of that is transportation, which includes cars and trains – but the biggest emitter by far is air travel.

You might be reading this and have never taken a flight and never intend to, or you might prefer to road trip and use trains for your summer holidays. If so, that is amazing, because you can save up to 90 per cent on carbon emissions by avoiding planes.

Flights currently account for 2.5 per cent of global carbon dioxide production, and because planes fly closer to the atmosphere, they create more damage than transport on the ground. This is a figure that the industry itself accepts, but, in reality, scientists always point out that it is only half the truth, as aviation emissions such as nitrogen oxides, water vapour and particulates have additional warming effects.

It is the mixture of all of the above that makes aeroplanes so polluting. Although regulators attempt to put pressure on airlines to reduce their impact, the demand for air travel is rapidly growing. On a macro scale, the air industry plays a huge role in ensuring the transportation of goods. Global supply chains, from food to the transportation of medicine and for aid, hugely depend on aeroplanes as one of the quickest means of transportation. Air transport carries about 0.5 per cent of the volume of world trade shipments, but it's over 35 per cent by value – meaning that goods by air are very high-value commodities, such as perishable food or time-sensitive items. For both personal and commercial uses, over 65 million jobs are supported in aviation and the related tourism industry. Of this, more than 10 million people work directly in the aviation industry[10] and thus this is not an industry that will be disappearing anytime soon.

Can travel be environmentally friendly?

As awareness grows of the need to reduce our individual carbon footprints in order to stop climate catastrophes, the aerospace industry is under sustained pressure to find clean solutions. For travel to become more environmentally friendly there must be tougher regulations within the private sector, because advancements in the aero sector are relatively low and hugely costly. The jet aircraft functioning today are, however, well over 80 per cent more fuel efficient per seat kilometre than the first jets in the 1960s. Although we can't wait another 60 years for such progress, there must be faith in humanity and those working in the industry that innovation and progression are very much possible. Companies are banking on a new generation of less polluting planes. New technologies, such as electric planes and the use of biofuels, which could reduce the environmental impact of flying, are in development, but it is a relatively slow process and lots of safety testing is required. Some airlines are trying to increase efficiency in the short term by reducing the weight that each flight carries, and by taking more passengers and reducing distances. Time is not on the side of the aviation industries and, so far, some of the 'solutions' only solve part of the problem and air travel is notoriously difficult to decarbonise.

As tensions rise between air travel demand and mounting climate change concerns, creating more uncertainty in the industry, the United Nations is working on setting up an emissions trading scheme for airlines to offset their contribution to climate change on a large scale. Recently, the heads of big airline companies have come out and openly acknowledged the challenges and admitted that the industry is not moving at the pace it should be. One thing is for sure, the argument needs to be made that while change is needed from a moral perspective, it can also provide business opportunities. If we can somehow prove greener travel is better for an airline's bottom line, we might just witness significant advancements in our own lifetime.

Planes do connect people to places and possibilities, and there are so many incredible benefits from seeing other parts of the world and learning about different cultures. Do we really need to travel as much as we do, though, or is the amount we fly induced by the industry promoting a lifestyle? In addition to the artificially low aeroplane ticket prices, are our flight choices a symbol of wealthy frivolous consumption? If you fly for pleasure or business, it is important to recognise the sheer magnitude of the impact that even just one flight can have on our planet.

- Eight hundred and ninety-five million tonnes of CO2 were produced by flights worldwide in 2018. Globally, all humans produced just over 42 billion tonnes of CO2 in the same year.
- In 2018, nearly 4.4 billion passengers were carried by the world's airlines.
- Just one single journey from New York to London on an aeroplane is said to melt up to 3sq.m of Arctic ice.
- Environmental group, German watch, estimated that a single person taking one round trip from Germany to

the Caribbean produces the same amount of damaging emissions as 80 residents of Tanzania do in an entire year – about 4 tonnes of CO2.
- In 1945, it took 130 weeks for a person earning the average Australian wage to earn enough for the lowest Sydney to London return airfare. In 2009, it took just 1.7 weeks.[11]

There is another crazy statistic that states that only 6 per cent of the world's population has ever flown, and with economic growth and globalisation, the number is likely to rapidly increase. According to studies by banking super-giant UBS, global air travel is growing between 4 per cent and 5 per cent every year, meaning that the overall numbers are doubling every 15 years[12] and thus the rising demand for flights is surpassing any reduction in emissions. But surely we can't reduce emissions by denying people who have never flown before the opportunity to fly? Also, how do we deal with proposed airport expansions around the world that are being strenuously opposed by climate activists?

Can you change your habits?

There are the extremes of the likes of Greta Thunberg who travel country to country only by train wherever possible. Or, when it came to crossing the Atlantic to attend the climate summit in New York and later on in Chile in September 2019, Greta hitched a wild two-week zero-emission journey on a 60-foot, carbon-neutral racing yacht.

Can examples like Greta trigger you to change your flying habits or even to challenge yourself to not fly at all?

In the end, if we can't change the bigger system, travel decisions come down to individual choice. On a personal level, there are ways, albeit not perfect, that are certainly positive contributions you can make, from carbon offsetting (see page 172) to reducing waste during your journeys by avoiding buying food and other items on the plane, which are always wrapped in plastic, as I discuss later.

It is important to note that short-haul flights under 500km are comparatively the worst polluters, because they take a lot of energy to get up into the air and down again. A return trip from London to Paris, for example, would produce 110kg of CO2 by plane vs 6.6kg by train. Shorter journeys are the easiest to switch for train or car and can be an area to potentially reduce your carbon footprint.

On an aeroplane you are quite literally in a tube of metal and plastic. Everything is wrapped in plastic, from the blanket to the cutlery and food – and even the vanity kit on a long-haul flight. This is needed to reduce germs in the confined space of the plane with air that is recycled.

If or when you next take a plane, take a couple of minutes to look around and observe the amount of waste consumed by fellow passengers. From plastic water bottles, serving cups, plastic utensils, drink mixers, food packaging – it is enough to give you a full-on anxiety attack. Travelling is like any other aspect of life, so just apply the tips you have read in this book to your journey. Here are some reminders:

- Take fewer flights.
- Reduce your flight stopovers wherever possible, and fly direct.
- If flying for work and business, are more Skype calls a possibility? Of course, sometimes nothing can beat face-to-face interaction with fellow colleagues and for project oversights.
- Take trains or ground transport over air, especially for those shorter journeys.
- Use a digital boarding pass when possible and avoid a printed boarding pass.
- Ditch the overpriced unfriendly airport toiletry minis – which are the epitome of single-use ridicule – in favour of reusable decanters and stackers to carry the right portions of your own can't-live-without products from home.
- If you are having trouble squeezing all your toiletries into the clear plastic bag they give you at airport security, swap products for solid versions. Buy your solids in everything from shampoo, conditioner, powder-compact perfumes and deodorant sticks.
- Travel with your reusable water bottle: most airports have refill stations and so do some planes, so don't hesitate to ask your air steward to fill it up.
- Travel with your other reusable items, such as coffee cups and reusable utensils.
- Avoid opening any flight toiletry bags, and bring your own products to use.
- Refuse as many in-flight plastic items as possible. Planes,

and especially economy flights, are rife with plastics. Say no.

- Bring your own food and eat vegetarian/vegan meals on board (remember that often these days you are asked not to eat nuts if there is someone with a nut allergy on the plane).
- If you take a longish flight, you are usually given a blanket and pillow wrapped in plastic – avoid when possible.

Carbon offsetting

While we, as individuals, are unable to control the pace of the entire air industry, one of the easiest ways to reduce your flight's impact on the environment is by participating in voluntary carbon offsetting schemes. People can contribute to offset or make up for the emissions that their flight or other transport produces by paying schemes usually to either plant trees or support green energy projects. This essentially somewhat neutralises the effect of your emissions.

Of course, offsetting your carbon does not confront the morality of air travel, and having this option may make you less likely to change your behaviour. Some people argue that offsetting is a luxury because it costs money and not everyone can afford to do so, and it completely overlooks the deeper structural issues of carbon emissions. Carbon emissions include the escalating temperatures we refer to as global warming, as well as extreme weather patterns, which destroys wildlife and habitats, contributes to rising sea levels and a magnitude of other impacts.

Of course, if we were not emitting as much in the first place, we wouldn't need carbon offsetting projects. Offsetting doesn't remove the carbon dioxide produced when you fly. That, of course, still goes straight into the atmosphere, but schemes like these can reduce the CO2 somewhere else in the world. For those of us required to travel by airplane, carbon offsetting is a cost-effective economically efficient manner of taking responsibility for our residual emissions.

Know your flight's carbon footprint

When you take a flight, whether you offset or not, carbon dioxide is going to be released into the atmosphere. Offsetting is better than doing nothing. It sounds great in practice, but not many people are using carbon offsetting schemes because they simply don't know about them or are dubious about which schemes to use. Here's how to offset:

- Step 1: calculate your emissions using an online calculator.
- Step 2: choose an offset project from the online portfolios.
- Step 3: start offsetting your emissions.

Offsetting schemes are usually based on measuring how many tonnes of CO2 are produced by each flight you take, through carbon calculators that are available online, and then by investing in a project that reduces CO2 levels by the same amount.[13] For most of us, calculating such a thing sounds completely alien and confusing, so various calculators have been developed to work out how much carbon dioxide you produce on a flight. The most basic metrics are how many miles you've flown, although others might take into consideration your class of travel, the type of aircraft and how fuel efficient it is, the number of passengers and the occupancy rate, and perhaps even the time of day and the weather while flying. Every carbon offset scheme calculates things differently. If you take lots of flights, it might be easier to create a spreadsheet to keep track of your flights over a certain period, such as every few months, then to offset them all at once.

Are you wondering why the class you travel matters when you are on a flight? In business or first class, unlike economy, there are significantly fewer passengers, meaning that there is a less efficient use of space.

Which scheme should you use?

Because the calculators of different companies and organisations work differently, the amount you can pay varies too, and there can be confusion over the different prices charged by different companies for the same activity.

If you are considering paying to offset your flight CO2, who the bloody hell should you offset with? There is a minefield of companies online with lots of impressive claims and websites covered in lush forests and pristine running streams, but not all are regulated and transparent with their funding. Unearthing the good from the bad is not easy, but here are some that are recommended by the Gold Standard organisation (which audits projects according to the rules laid out in the Kyoto protocol), which has taken the time to investigate and certify the best. Always make sure to select an individual project to fund rather than a company portfolio so that you know where your money is going, and you can feel like you are personally getting the maximum benefit. Some of the best offset CO2 schemes have long since switched to energy projects – anything from distributing efficient cooking stoves to capturing methane gas at landfill sites. Energy-based projects such as these are designed to make more permanent savings than planting trees; they also help poor families to save money on fuel and improve their household air quality.

The easiest way, however, is to offset directly with your airline when you book your flight. About a third of airlines allow you to pay an extra fee on top of the flight cost, which is donated either to their own offset programme or someone else's. Recommended offsets favour energy projects over tree planting, for example:

Climate Care (www.climatecare.org) This is a company that has ethics and people deeply rooted in all their nature practices. They have a great online calculator to help total your carbon-travel footprint, not only for flights but also for car, home energy, events and businesses, too.

Gold Standard (www.goldstandard.org) are considered to have one of the most rigorous climate standards by many respected NGOs globally. This means that every pound spent offsetting with them not only offsets CO2 – there are 1,700-plus projects across 80 countries in their portfolio to choose from – each project creates value for local communities and ecosystems, and contributes in a measurable way to the sustainable development goals.

If you work for a company that has employees who travel frequently, encourage your business to start an offsetting scheme that can be part of the company's corporate social responsibility scheme (CSR); it will look good for their profile too.

PERSONAL CALCULATIONS USING CLIMATE CARE

I couldn't write this without being fully transparent about my own travel path, and those who know me personally can attest to my significant carbon flight footprint. From a young age, the opportunity to travel opened up my horizons, and provided incredible experiences that completely shaped me into who I am as a person. Today, my time in the air is predominantly for work and business, something that unless I decide to change jobs will remain much higher than the average person.

Here are some of the frequent return routes that I take, mostly for work. I calculate my carbon emissions on the ClimateCare.org website:

▶

London to Madrid return:

0.39 tonnes of CO_2
1,546.47 miles travelled
£2.91 cost to offset

London to Ghana return:

1.53 tonnes of CO_2
6,339.05 miles travelled
£11.46 cost to offset

London to Kenya return:

2.05 tonnes of CO_2
8,502.55 miles travelled
£15.37 cost to offset

The total cost to offset three of my main travel routes is £29.74. I choose to support grass roots projects that deliver value for communities in developing countries as well as the environment, backed up with verified offset certificate and carbon credits. Some favourite projects include LifeStraw Carbon for Water, supplying people in Kenya with safe drinking water, and Gyapa Stoves in Ghana, which are efficient cooking stoves that reduce harmful smoke emissions.

6

Futureproof: Social Media and Business with Purpose

> 'Imagine if trees gave us WiFi signals, we would be planting so many trees we'd probably save the planet too. Too bad they only produce the oxygen we breathe'
>
> Tarun Sarathe

Social media is an incredible and informative tool, and it has revolutionised entire industries, in particular marketing, and the way brands interact with consumers. Over 3.4 billion of us across the world use social media regularly every day to eradicate the boundaries of time and distance, bringing like-minded people and businesses together across the globe. As a result, we are not only the most connected generation, but also the most informed.

There are no longer excuses for not knowing what is going

on in the world around us. We can use social media to not only learn about fundamental issues such as climate change but also to share our opinions, voices and support for our beliefs. But, intertwined in the power of social-media knowledge is also a mental parasite that creates a world where no material resource feels enough, and as we strive for more we not only consume ourselves but our planet, too. Over half the world is now using social media daily, and 54 per cent of those use social media to research products and to shop. Ninety per cent of those purchasers are millennials and the Gen-Z generation. How we feel on the inside is shaped by what we do on the outside, and our minds are products of our environment.

This chapter delves into the rise of social media, our consumer behaviour and what the future of businesses with purpose could look like. All this is echoed by the voices of a few incredible sustainable influencers and business pros, who have shared their thoughts with me for this chapter.

Social media and our modern world

There is no direct link between social media and our changing climate. Social media platforms have not necessarily contributed to greenhouse gas emissions, the plastic crisis or the rise of fast fashion. But they are fundamental in our climate dialogue, because social media is not a movement for Gen-Z – it is embedded in our culture and fundamentally tied to many decisions that we make. Today the average Brit spends over 34 hours a week online on their smartphone scrolling, liking, sharing, swiping, skimming and responding to people's thoughts. Young people are so addicted that we check our phones on average every 8.6 minutes[1] and most of us say that we get our daily news from social media platforms. It's safe to say that social media single-handedly shapes our ideologies, beliefs and consumer behaviours, and it connects with our perception of social and scientific issues such as climate change.

An early and popular definition of social media said that it is 'an online structure where individuals use their own profiles to connect with other individuals by creating lists of friends' profiles.'[2] Yes, that was a definition from circa 2007. That is what Mark Zuckerberg had in mind when growing Facebook. The platform is still in many ways the same platform that thrives on people connecting – no one really logs into Facebook hoping to see a company's advertisement, but they will tolerate it if it's personalised and provides some value in the sense of authenticity.

Today, advertisements and narratives are being written and re-written every single day by those participating on social media. Through algorithms and artificial intelligence certain social platforms distribute streams of content that reflect our existing beliefs and political affiliations rather than scientifically accurate content that might question our biases. Certain companies even specialise in tapping into the dark side of social media masked under 'engagement', trying to hack our brains with fake news and inflammatory headlines. Our engagement determines the type of content that we see. This is perhaps when we, as individuals, need to reconsider the authenticity of our beliefs and the visual 'diet' we consume.

Environment and social media

The research behind online environmental communications is rare; the data is sporadic, continually growing, and the effectiveness of climate posts are little known. The ideal social media chapter in this book would be able to tell you that X amount of environmental and ocean organisations exist across Instagram and X amount on Facebook, their total reach and their effectiveness. Unfortunately, after hours of Googling, very few of the numbers found are concrete enough to quote, and most are opinion based.

On a more conventional note, the news is rather grim; there is a virtual black hole when it comes to green news, as major news outlets favour crime and entertainment stories above environmental ones. When it comes to climate change, big media players in the past have failed to report on the huge scientific findings and progressive UN reports and instead focused on the drama of the players involved in the negotiations. Climate change news is generally associated with negative content such as natural disasters, oil spills, ice caps melting or sea life dying from plastic ingestion. This is not to say that media outlets don't produce credible journalism on the issue, previously it didn't help that traditional mainstream media failed to substantiate facts on climate content, and different articles quoted distinctive statistics, potentially heightening the public's distrust in the climate change argument.

It could also be argued that mainstream media is afraid of making people scared, by narrating future apocalyptic scenarios that are an idea of what the future *might* look like. Although it's certain that environmentalism doesn't always snag the major headlines, we are now seeing, almost daily, important coverage of climate issues by news agencies. The media can no longer hide behind potential scenarios and ideals by scientists, instead natural disasters and large-scale climate catastrophes dominate the news. Climate change is our new reality.

Groups such as Extinction Rebellion have used the media to call out major governmental players to tell the truth about climate change. This, however, did not come without a struggle: at the very beginning of the group's existence, they accused the traditional media of staying silent on the ecological emergency for too long. Although both sides might not be able to agree on what should be reportable or not, the explosion of strikes globally initiated by groups such as Extinction Rebellion and youth leader Greta Thunberg saw significant improvements on climate coverage. Millions of people across the globe, and hundreds of thousands across the UK, have sent a clear message about the need for urgent action. The sheer scale of protests and vigorous activism became impossible for media outlets to ignore.

Politics can also interfere, and political parties pay media outlets to portray climate change in a certain manner or in so vague a way that it becomes too complex for the average person. The power of social media is that it provides a democratic platform for people to speak the truth. For instance, Indian media portray a nationalistic attitude towards climate change with the idea that compliance with carbon emissions reduction will limit India's economic growth. Alternatively, Japanese media found a positive relationship between the amount of coverage of climate change and public concern for it.[3] Yet sometimes the media actively ignores broadcasting the severity of climate change and covering environmental protests and carefully avoids talking about the forces and people behind the degradation of our planet. The idea that politicians and a vote will save us is not enough; action required to change the system cannot wait – 'they' won't, so *you* must get involved.

The headlines dominating reporting of protests such as those of Extinction Rebellion can often be misleading, insinuating this battle is only for governments to tackle. It is often emphasised that it's not up to the individual but the government and it is their burden to make changes. We must start reframing the climate discussions and scream and shout about how it's also up to us as individuals to change our lifestyles and consumption habits. By breaking down climate change on a personal level and covering this angle more, we can make this overwhelming topic understandable and relatable to the masses, encouraging and empowering more people to get into climate conversations at home and to reassess their contribution. The biggest mistake is thinking climate change is someone else's problem to deal with it. It's not, it is everyone's and ours to deal with and solve. We need to start portraying climate change as a problem everyone can understand, and how each individual is a contributor one way or another – and that it has recognisable personal costs for all of us.

Climate-friendly action through social media

The global domination of the minds of young people by social media is slightly terrifying, but at the same time social platforms are providing spaces for collective voices to share their opinions about the things that they care about. The #plasticfree and #zerowaste hashtags are some of the most popular online, with people sharing their tips on how to live a waste-free lifestyle. The rise of organised climate strikes, environmental posts and sustainable content is a testament that people are hungry for more climate news.

Although social media allows us to have collective power to mobilise the fight for climate-change solutions, and education, it can also be undermined by an influential individual or group with a different opinion. Finding the balance of how we use social media to its optimal potential for issues such as climate change is a path that current and future generations are figuring out. Although there are a lot of questions waiting to be answered as our digital lives continue to evolve, if we focus on the positive, social media might just be the modern tool our planet needs.

Social media is a place where individuals can find comfort in their collective tribal identities. By posting, liking and affirming certain content, we generate social value, regardless of what the subject might be. Sharing platforms encourages greater knowledge of climate change, assists the mobilisation of environmental activists, and is a key tool in communicating the topic for organisations fighting the cause. It must also be noted that social media has been a powerful tool following climate change-related disasters. These platforms have been vital in coordinating rescue and relief operations in the aftermath of such catastrophes.

Despite its drawbacks, social media gives us a platform to discuss, educate and debate climate change, so we need to use the modern digital tools available to us to start a dialogue with other citizens and decision-makers. This is where you as an individual can use your social voice to share accurate climate information with your family and social circles.

The power of influencers

Social media has been both good and bad for the climate movement, depending on what side of the fence you sit on. Certain individuals have the power to garner a great deal of online attention regarding what music they listen to, what they wear, and what they eat, say and do.

I assume that you follow at least a few social media influencers through various platforms. If you use Instagram, open it now and scroll down. It's likely that you won't go more than a few seconds without seeing someone you follow promoting a brand. These people are known as influencers.

An influencer is considered to be an individual who has the power to affect the purchase decisions of others because of his or her authority, knowledge, position or relationship with his or her audience. Today, content marketing for brands can be much cheaper than traditional marketing, depending on the influencer. Curating content online is a full-time occupation for thousands of people globally, and it is a job title that many adolescents heavily aspire to.

Whether you agree or not with the rise of the influencer – what he or she is, how they monetise social platforms to build business empires, and whether you trust them or not – their presence is incredibly powerful. Circa $5.6 billion were spent on Instagram influencer marketing in 2018. Brands also recognise that content marketing costs 62 per cent less than traditional marketing and generates, on average, three times as many leads.

The important question for us is how can influencers use their platforms for social good, such as raising awareness of climate change?

There are a rising number of bloggers who have started using their platforms to promote sustainable lifestyles, fashion and plant-based diets. They are openly advocating climate-change awareness through content creation and brand partnerships. Their power is not only educational, but it is also encouraging sustainable change across the masses, and, as a result, the knock-on effect has been brands aligning themselves with influencer viewpoints realising good is the new cool.

Some of my favourite sustainable influencers and organisations, who are creating inspiring content and using their social platforms for a good purpose, have shared their words below about how and why social media is a great tool for saving the planet.

Zanna Van Dijk
Fitness, food and conscious-living blogger, @Zannavandijk

Q How and why is social media a great tool for helping to save the planet and inspiring action?
A Every single person has the power to influence others on social media, whether you have 100 or 100,000 followers. By simply sharing what you're learning about the climate you never know who you could inspire to make a positive change. You might bring the subject to light for someone who had never even thought of it before, and by leading by example and showing sustainable swaps, what you're doing can empower others to join you. Honestly, the possibilities with social media are endless, and I encourage everyone to embrace sharing their sustainability journey online – even if it's just little nuggets here and there.

Q Why did you personally choose to use your own channels to spread meaningful messages?
A I am lucky enough to have a platform which a few hundred thousand people watch – a true privilege. And I don't want to take that for granted. I take my role as an influencer very seriously, and I want to be a positive influence, a blogger with a purpose. It gives me so much fulfilment to share my passion for the planet with my followers and see them join me on my journey. Even inspiring one person to reduce their meat consumption or plastic use brightens my day, and I am constantly chasing those little rays of sunshine.

Q How has channelling sustainability brought new opportunities your way from a business growth perspective?
A It has opened so many more doors than it has closed. I always thought that talking about sustainability would push brands away from me, but as society becomes more #woke, so do brands. It is incredible that I get to support them in their journey in trying to become friendlier to people and the planet. I can only hope that more brands jump on board this movement.

Bradley Davidson
Head of media at Ocean Generation,
@Oceangeneration

Q How is social media shaping the future of businesses and organisations, and how valuable is data in the digital sphere?
A Social media is increasingly utilised as a mobilisation tool. For both good and bad, varying demographics can be influenced to think, shop and even vote in a variety of ways. Social media blurs the boundaries of freedom of speech and is a catalyst for propaganda. As worry increases about those holding power, we should focus entirely on the positive effects this tool has on society, in specific regard to the environment. We know that *National Geographic* is one of the most followed social media accounts in the world, with an Instagram following of 127 million. So, what does this say? It means that this company has a wealth of analytics on the most engaged environmentalists around the world. It means that if this data was able to be bought by non-profits and

environmental projects they could propel their call to action. Data is now considered more valuable than oil – ironic in the context of climate change. So should we, as social media users, protest that if tech giants continue to exploit our data they should at least donate it to good non-profits for free? Especially when traditional marketing strategies are blocked by social media companies if an organisation applies to advertise any climate change related content. Political propaganda of any magnitude or message can, one way or another, be advertised through Facebook, yet, if an official environmental organisation tried to advertise their call-to-action(s) or any related content, they face immediate sanctions, often without appeal. This must change.

As we move into this unknown sphere of digital human evolution, data is the most valuable asset of all. But, as social media users are starting to learn, our data is already being exploited beyond our imagination. We should at least encourage our data to be donated to good causes. We have a choice when we use social media to connect the dots and shift the waves of change. Social media is just the vehicle for data and data has the power to change the world.

Doina Ciobanu
Creative and sustainability advocate, @Doina

Q How and why is social media a great tool for helping to save the planet and inspiring action?

A Social media connects all of us in the most magnified manner.

Unlike other connection channels such as texts, calls, news and so on, the content of social media is so easy to consume, and that is where we have a unique chance of communicating the sustainability message in an easily understandable way. I think that is the key. The sustainable and environmental messages must be simple enough for everyone to connect with. Sometimes that might also involve imagery and easy steps to take.

Q Why did you personally choose to use your own channels to spread meaningful messages?

A Because if not me, who else? And I think that's how we should all think about sustainability, both when making little adjustments in our life but also when advocating for it. I approach it from my very personal perspective. I feel the responsibility to talk about it because there is no time to expect others to do it on my behalf. I need to do my part, and spreading awareness is the least I can do.

Q How has channelling sustainability brought new opportunities your way from a business-growth perspective?

A In terms of business opportunities I started consulting to help businesses become more sustainable. However, aside from the business side, what is most important to me is the number of my followers who are constantly writing to me about having adjusted their individual lifestyles, off the back of my content, to be more sustainable on a daily basis through their choices. That is the most phenomenal outcome I could ever have hoped for – using my social media platform for inspiring action. Changing even just one person's one unsustainable habit into a more sustainable one is exactly why I do what I do.

Extinction Rebellion (ER) spokesperson, @extinctionrebellion

The movement, Extinction Rebellion, was born out of the premise that this is an emergency as climate breakdown threatens life on Earth. Our politicians have failed us, so we must rebel. Extinction Rebellion is thrusting the topic on to people in city centres and across social media, making it harder for people to ignore responsibility. Through non-violent and disruptive action, the groups are using daring tactics to grab attention and overthrow the current systems. The scale of the protests and media domination would not have been possible without the use of awareness, organisational logistics and calls to action across social media.

Q Why did ER choose to start with social media channels to spread its meaningful message?

A We use social media because it allows us to share scientific facts through artistic forms, such as photography and graphic design. It allows us to share the truth about the climate and ecological emergency in ways that people might not see otherwise. It is particularly useful in reaching out to young people and global communities. In showing that we are fighting for what we believe in, we aim to inspire others to act with us. By providing credible scientific findings in visually appealing and gripping ways, and showing images of our volunteers in action, we are leading by example: social media lets us show the dedication, determination and successes of our actions and helps us to inspire others to join the cause.

Q How and why is social media a great tool for helping ER to promote its messages and inspiring audiences into acting for the environment?

A In this day and age, social media is an invaluable tool for raising awareness and inspiring individuals. At Extinction Rebellion we utilise multiple channels of media – Facebook, Instagram, Twitter, YouTube, podcasts, and so on – in order to share the truth about the climate and ecological emergency with as many people as possible. Through posts and stories, we can reach a wide demographic of individuals and offer the public educational, motivational material concerning the state of the crisis. Social media gives us the potential to educate and empower anyone – and everyone – to act in the fight for our futures and the health of our planet.

Q How has channelling climate activism brought new opportunities for the group and been a catalyst for change at higher levels other than grassroots?

A Social media has allowed us to connect with thousands of people around the world. It has given others a point of contact and helped to inspire and create Extinction Rebellion groups in different communities and countries; it has connected us with valuable celebrity spokespeople and incredibly talented artists. Extinction Rebellion's social media pages have reached the masses in ways that we couldn't if it didn't exist. It has digitally connected this incredible community, allowing would-be strangers to inspire and elicit action among their peers and communities. In these ways, social media has been an incredible tool and driving force in helping our messages spread, our actions have impact, and our movement grow into a global phenomenon – all in the name of a global climate and ecological emergency. Are you ready to join the rebellion?

Q Is there anything else you think might be relevant to the social/climate conversation?

A Social media is motivating people around the world to get creative and act on the climate and ecological emergency. To know we have activists such as Bhavreen Kandhari in Delhi, who helped set up ER India after learning about the movement online and tweeting about it in October 2018, makes us realise how powerful these tools can be. Social media is helping us to get the word out about this emergency into homes all around the world. And it's the whole world that needs to hear this message if we're going to avert the greatest crisis humanity has ever faced.

The damaging effects of denialism

I have talked about the power of good and raising awareness about the severity of climate change, but it wouldn't be fair to only share the positive voices out there. Science and theories can easily be discarded or discredited at the click of a finger.

The first draft of this chapter contained some incredible, unfathomable and outright obscene examples of tweets from some climate change deniers; however, due to publishing and copyright laws those unfortunately didn't make it past the first draft and it makes this chapter a little less comical. The point is, powerful people with large followings across social media have the dominance to make false claims about the environment and spread fake news and fake science about this very serious topic. As polarising and divisive as political or celebrity figures can be, whenever they speak about climate change, immediately they send the online world into a state of frenzy.

One does have to question where they get their ideologies and 'science' sources from, but we must hope that when these statements go viral, the audience's attention goes in the opposite direction, raising more awareness about climate change in general, which in reality is a win.

The reality of fake news shows how social networks can be fragile in important ways. It is hard to know what a comprehensive solution might be; not even the tech giants themselves know what to do. Social media platforms such as Twitter, Facebook and LinkedIn allow almost anyone to publish their thoughts or share stories to the world, and most people don't check the source of news before they share it, which can lead to fake news spreading quickly or even going viral. Platforms such as Facebook are attempting to launch specific news feeds with regulated content, using algorithms to fact check and approve.

At the end of the day, though, the social media platforms were created to connect people, and now their responsibilities are shifting into policing the world. Although we must hold social media platforms responsible for serving misinformation, we can also help them to do better. Even though it is becoming increasingly difficult to identify sources of news stories and assess their accuracy, we should take the moral responsibility of doing so. We can build immunity to misinformation by exposing people, reasoning fallacies in denialist claims, and turning around the untrue hype, to garner more support for the climate conversation.

How effective has social media been?

It varies hugely what type of content garners our attention and support across social media platforms. From popular culture to fashion to sports, there is social media content to pique curiosity in anyone. In recent years the spread of other information outside of our personal preferences across digital platforms has been a powerful and far-reaching tool to spread awareness and educate people about politics and climate change, sparking lots of individuals to question what the hell is going on in the world.

But, over and over again, the impact of climate news has led to very little change in our daily lives. Why? Is it because people today are overloaded with information and distracted by negativity, resulting in most of us failing to act to save our planet? People who glaze over the call for action clearly believe someone else will come to our rescue, ignoring the chaos. Perhaps some of us even throw our hands in the air, saying that it is too late to act and relieving ourselves of the moral urgency.

We are living in a time where the world is full of negativity, and it's easier for us to just scroll through meaningless content. As I have discussed before, climate change is an incredibly abstract topic for most people, and although everyone generally has some opinion on the topic, it is very difficult to measure how it translates into 'offline' action in support of the cause. Non-believers may argue that if there is no offline action, why bother using social platforms for climate change? Social media and other online mediums, such as gaming and blogging, are more opportunities for climate education and spreading knowledge.

AN EXAMPLE OF HOW WE CAN USE SOCIAL MEDIA FOR GOOD

In 2017, non-profit organisation Ocean Generation teamed up with global game developer Rovio, to incorporate ocean messaging into one of the most downloaded game franchises in history – Angry Birds. To communicate with Generation Z in a dynamic way, Ocean Generation launched three mini games: The Big Catch, about plastic pollution; Small Island Defence,

▶

about rising sea levels; and The Last Straw, which is about the plastic straw epidemic. Throughout the experiences, players encountered fun game mechanics while being educated and presented with clear calls to action about the oceans. Once the games were completed, players were directed through to other ocean resources and invited to make personal commitments with regards to their use of damaging plastics and individual action.

During the 3-month partnership between Ocean Generation and Rovio, over 7 million people played the games and afterwards over 300,000 pledged, across various platforms, to change their daily habits. This is just one example, ultimately proving that digital social platforms can educate and inspire behavioural change.

Scholars recognise that public opinion forms more readily when events are psychologically closer to the individual. Those battling at the forefront of climate change every day believe in the climate facts and don't need an online post to reaffirm that, but only a limited number of people are in that position. Those who do not see or feel the consequences of climate change on a frequent basis are, of course, less likely to care, but social media has the potential to personalise social issues in many ways and is thus an appropriate lens through which to alter opinions on climate change for the better. Would telling the stories of people in distress because of climate change help to spur you into action, for example by reducing your energy consumption? Has the viral video of a turtle with plastic stuck up its nose and someone trying to get it out influenced you to change your single-use plastic use?

Nature has an incredible way of fighting back and regenerating what was lost by human destruction. The task of using social media for better results is therefore much less daunting than we might imagine. Erica Chenoweth, an American political scientist, reveals that for any peaceful mass movement to succeed, a maximum of 3.5 per cent of the population needs to mobilise. Humans are ultra-social mammals, constantly, if subliminally, aware of shifting social currents. Once we perceive that the status quo has changed, we flip from support for one state of being to support for another. When a committed and vocal 3.5 per cent unites behind the demand for a new system, the social avalanche that follows becomes irresistible.[4]

The climate is changing – you can too

What can you do with your voice? Be part of that 3.5 per cent and use your voice in the following ways to help raise awareness and inspire behavioural changes for the planet:

Do not underestimate the power of your voice across social media and your ability to share climate knowledge with family and peers.

Celebrate Earth Days such as on 22 April and World Oceans Day on 8 June. These days are extra-special chances for everyone to celebrate across their personal social channels and to reconsider how our decisions affect the world around us.

Never underestimate the power of liking, sharing and posting about your favourite organisations online. For those who create climate-awareness content your engagement makes it worthwhile and it can lead to further opportunities for collaboration and greater change.

Brands are beginning to understand that people want to share unique experiences. Team up with brands or organisations such as Surfers Against Sewage, Ren Skincare and Parley, which are organising beach clean-ups globally, or search Eventbrite in your local area for sustainable related events. Go to these events, actively be involved with sustainability and share them with your network across socials.

Follow influencers with positive agendas who are inspiring change and encouraging sustainable lifestyles.

Follow and engage with planet-based bloggers for veggie, vegan chef inspiration. One of the easiest ways to reduce your personal carbon footprint is to reduce your meat consumption, as we saw in Chapter 4. Lots of these environmental activists have delicious recipes that are bound to satisfy your taste buds any day of the week.

Sign online petitions Adding your voice to the causes that resonate with you sends a strong, political message that you care, and it is so easy to do with just a few clicks online.

Make Ecosia your new search engine and plant trees with your searches for free. Add the Ecosia tab to your browser extension. Download their free browser extension and as you search the web with their platform, the search ads generate income that Ecosia uses to plant trees. To date, Ecosia has over 8 million active users and as a result has planted over 80 million trees around the world.

> 'As consumers we have so much power to change the world by just being careful in what we buy'
>
> **Emma Watson**

The future of business

Approximately 1.7 billion people worldwide belong to the 'consumer class' – the group of people characterised by diets of highly processed food, the desire for bigger houses, more and bigger cars, higher levels of debt, and a lifestyle devoted to the accumulation of non-essential goods.[5] It is not the billions living in poverty in towns and villages of sub-Saharan Africa or the Far East who are indulging in the rapacious consumption that is breaking the planet. The poorest 50 per cent of the world's population is, according to the charity Oxfam, responsible for a meagre 10 per cent of 'lifestyle consumption' carbon emissions. Europe and America, for example, with about 12 per cent of the global population, account for over 60 per cent of worldwide consumption.[6] However, nearly half of global consumers reside not only in Western countries but also in developing countries too, including predominantly 240 million in China and 120 million in India – the markets with the most potential for expansion.

It's easy to hate on consumption, and way more of it happens in rich countries. Consumerism has poisoned areas of contemporary life, and, together with commercialisation, it is a major driver of climate change. Our need for instant gratification through purchasing has led businesses and individuals to become incredibly powerful and rich off the back of our behaviour. The goal of quick profit and limitless growth has been the foundation of many businesses, maximising shareholder profits no matter what the human or environmental cost.

Globalisation – the spread of consumerism

As we move into a new space in human civilisation, the real threat of climate change is beginning to affect every single living organism on our planet. It could be described as an experiment gone wrong: let's just see how far we can push every eco-system past the point of repair; let's chop down the lungs of the Earth that feed us oxygen; let's clog up our oceans with trash, poison our drinking water and pollute the very air we need to breathe to survive in order to buy, buy, buy.

This rising consumption has helped to meet basic needs, create jobs and essentially is what our capitalistic society relies on. Globalisation is the driving force that makes goods and services that were previously out of reach in the developing countries much more available to them. Items that were once considered a luxury are now mainstream around the world, and our obsession and need for more, more, more has led us to this point of potential no return. The increase in prosperity does not necessarily mean a happier and healthier life, however. Rather, increased consumerism can come at a steep price for some, such as families who are incurring debt and working longer hours, sacrificing their lifestyle and community priorities to afford to buy more.

The reasons for this mad experiment could be considered strange if you were to change your perception slightly. Most of the global issues suffocating our planet essentially fall into four categories of consumption:

1. The process of moving something or someone from A to B.
2. The creation or manufacturing of commodities that are then moved from A to B.
3. The food we eat.
4. The clothes we wear.

Just years away from global catastrophe, organisations and governments continue to battle with the four categories above and the finance that connects them. The dilemma is simple yet tricky, all four pillars prop up the global economy, the same economy that feeds and clothes us. Yet to change the course of our future, these four pillars need to be dismantled and rebuilt with new science, innovation and research. On an individual level, conscious consumerism means making positive decisions throughout our own buying process, with the intention of helping to balance some of the negative impacts that consumerism has on the planet. Take responsibility for your impact, because remember that:

> 'Every dollar you spend ... or don't spend ... is a vote you cast for the world you want to live in'
>
> L.N. Smith

THE BIGGEST PROBLEMS

- Every day, in the US alone, e-commerce packages travel about the same distance as going to the moon and back, not just once but equivalent to 133,000 times.
- Seventy-one per cent of greenhouse gas emissions are produced by 100 companies – all of them are oil and gas companies.
- Twenty oil firms are behind more than one-third of all CO_2 emissions.

▶

- These companies knew how harmful fossil fuels were back in the 1950s but they carried on anyway. Some of the oil companies have gone so far as to publicly say that they are not directly responsible for emissions, since consumers are using the fossil fuels.

> 'The corporations and industries that we are up against will not go down without a fight. But they will not win, they will never win, because for all of their money and for all of their political influence, we have one thing they will never have. We have the truth'
>
> **Ed Winters, vegan educator and content creator**

Conscious consumerism

What if we approach this dilemma from a different angle – an angle that can work hand in hand with consumerism? The ambition is simple yet overlooked. Between the issues affecting the world around us and our immediate future, there are numerous networks of habits, passions, beliefs and politics already built into better business practices and what could otherwise be described as conscious consumerism. Although there is a lot of hype around socially conscious behaviour en vogue, the reality is that the conscious consumer spending index data are not exactly in tune with our desires. Part of the reality is that there are not enough socially responsible products – or services practising green behaviour and championing philanthropy – for us to choose from to satisfy our needs.

Our generation is reimagining our legacy, in order to demand better so that we can put our money where our mouths are to buy more consciously. People such as Greta Thunberg and other activists represent real hope. They all pose challenges for companies and how they engage with consumers. The new mobilising individuals and groups represent impetus, are unambiguous in their demands and will not be steeped in the traditional corporate engagement model. New activists are changing business-as-usual practices.[7] Companies greenwashing or NGOs funded by unethical corporations are being exposed across social media.

New movements and individuals in the spotlight are untamed beasts across social media, and if companies choose to ignore the likes of Greta, this generation have wallets, and soon, in the case of the young climate strikers, they will have votes. The more human a company's response to activists is, the better placed they will be to build resilience to the planetary turbulence now upon us, and brand loyalty. The priorities of the younger generation and upcoming consumers are not about leaving a house and inheritance to our kids. Instead, our priorities are ensuring that our kids have food and water available. There is, of course, no easy solution to encouraging conscious consumerism on the buyer's side, nor enforcing greener and ethical practices into businesses.

One man on a mission to solve one of the biggest problems facing companies and individuals in the 21st century is Afdhel Aziz. He helps companies, both old and new, around the world to unlock the power of business to do good in the world, helping them to find purpose and meaning in the work that they do. His journey has led him to advise important Fortune 500 companies and he is co-author of a book, *Good is the New Cool: Market Like You Give a Damn*.

Here is an interview with Afdhel about what conscious consumerism is, what the future of business looks like and how brands can be catalysts for the green movement.

Afdhel Aziz, @afdhel

Q When did you realise that every business with a purpose was a powerful case for the future of people and planet?

A I think I first started seeing it in 2011 when I started writing the book with my co-author Bobby Jones. What is astounding to see is how fast this idea of business is evolving to a 'stakeholder' capitalism model (as opposed to traditional 'shareholder' capitalism). The kinds of companies our social impact consultancy Conspiracy of Love are working with run the gamut of every category: lifestyle companies such as Adidas, Sonos and Bombay Sapphire Gin; technology companies such as Microsoft and Facebook; CPG companies such as Coty, Mars, Nestlé and Mondeléz – every category leader is exploring this idea of purpose-led business. However, I want to sound a warning sign. We are not moving fast enough to fix the problems caused by rampant consumerism. We are generating more carbon emissions and more waste than ever before. The climate crisis should be top of the mind for every single business leader as a problem that needs fixing. As Patagonia founder Yvon Chouinard said, 'We cannot do business on a dead planet.'

Q What does your idea of Good is the New Cool mean to you and those brands that you work with?
A Our initial thesis was that brands had to start thinking about 'doing good' as much as they thought about 'being cool'. In fact, we'd argue that doing good is its own form of cool. In increasingly commoditised categories, with little differentiation on product and prices, consumers are now 'voting with their wallets' and evaluating brands on their environmental and social credentials. So, it's essential that brands evolve to stay relevant; however, when we say 'the new cool' we certainly don't think that it's a short-term trend that will disappear: we believe that because it's built on new expectations from two new generations of consumers, millennials and Gen-Z, who have enormous spending power, we think it's here to stay. It's a new paradigm for business.

Q What are some of the best examples (climate related, if possible) of brands using the power of good not only to make the world a better place but also to see sustainability create a virtuous cycle of profitability too?
A I think the work that IKEA is doing around transforming itself

for the circular economy is fascinating. They are rethinking every single bit of their process: from supply chain and manufacturing to repair and resale.
Adidas is also an amazing brand that has proved that sustainability can be profitable: their Parley Ocean Plastic shoes are on track to sell 11 million pairs this year, generating more than $2 billion in revenue.

Q What does the future look like for an ideal world filled with conscious consumerism?
A I believe that it's essential that every business shifts to a circular economy model as fast as possible; that is the only way that we can continue to be a consumption-driven culture without damaging the planet irrevocably. Every business should be looking to create a model that has a zero-waste, zero-carbon, regenerative model at its core. We have the tools and technology available to make that shift. All that is missing is our collective will.

A new way forward for businesses

Pope Francis said, 'in the end, a world of exacerbated consumption is at the same time a world which mistreats life in all its forms'. It can be easy, as I have discussed throughout the book, to highlight the role of us as individuals to be conscious consumers, but businesses play a huge role in ensuring that we have sustainable and ethical options readily available. Business must recognise its role and responsibility within the system.

Their efforts must promote a new way of thinking and solving problems, and their relationship with nature. Conscious consumerism will also require us to view the future with 'attitude' innovation, such as the abandonment of us owning all our own things, such as cars or homes, in favour of the sharing economy.

Previously, people have called for green taxes to reflect the true environmental costs of a product. Other businesses, off their own backs, have started to recycle packaging or goods, or to run consumer education and awareness programmes. In providing future security, the global population faces many challenges, since our resources are already depleting with little room for compromise. Fundamentally, and foremost, we need to rethink the way of brands and businesses. The goal is not to focus on businesses selling less but on how to provide a better quality of life for all humans on the planet. For brands with purpose, leading the way, transforming the way we produce goods and thus the way we consume them, the key to success will be efficiency.

Obviously, it costs more to produce ethical, high-quality products that give back to people and the planet. Studies have confirmed that consumers will pay more for socially responsible products, but there is a limit as to how far we can allow prices to rise. We need to move away from the idea that 'do good' products are luxuries – they need to be the norm. Business has the power to unpick the labelled stereotype of consumers who are assumed to shop consciously. In recent history, the narrative is guilty of categorising conscious consumers as a movement driven by millennials or Gen-Z (and yes, we are the champions of demanding better from businesses and governments) *but*, in reality, this is not effective, because not everyone in this generation has the required purchasing power. Instead, we have to look to and incorporate the values of different demographics if we want to identify and expand conscious consumerism around the world and make it completely accessible across every industry.

Technological innovations

We live in the most advanced time in technological innovation. Unfortunately, though, early technological advancements, and those throughout history, even some that we still use today, have caused considerable damage to our environment. As we try to clean up the damage caused by previous technological innovations, we must use this space for good. When used with the environment in mind, technological advancements will be one of the most powerful tools to help us transition into a more sustainable future.

The rise of technology and biotechnology, although still in their infancy, will create real-world solutions to climate change. We are waking up to the potential of Blockchain and Bitcoin to create a different technological and financial infrastructure on which to build new enterprises, in a way that democratises opportunity for the masses.

It is a fascinating space to watch, as solutions driven by technology will bring together governments, research institutions and billionaire investors who are trying to limit climate change. So far in trying to achieve that goal, areas such as energy are trying to figure out safer and cost-efficient ways to create commercially viable green energy sources.

Humans need energy but don't need fossil fuels. We can, in fact, survive and exist in harmony with green energy. We have already found alternative ways of powering vehicles, such as with electricity, but in order to scale we need much more efficient batteries and charging technologies. The future of self-driving technology is in electric cars. In ensuring food security and reducing our dependency on animals, companies such as Beyond Meat are producing lab-grown meat – alternative meat substitutes that look, taste and feel like the real thing. What we need is for advancements like these to reach the market as soon as possible – we don't have time to wait.

High standards in business – B Corp

B Corporations are businesses that meet the highest standards of verified social and environmental performance, public transparency and legal accountability to balance profit and purpose. On the simplest level, B Corp certification is a stamp of approval for companies that have proven their commitment to doing good, verified by the founders of B Lab, a non-profit organisation. In essence, a B Corp inspires us to use the power of markets to solve social and environmental problems with no room for greenwashing. The B Corp really is the golden standard certification of a company, globally acknowledged and even known as an attractive addition for investors.

Started by three entrepreneurs, B Lab has spent over a decade championing the idea of a third-party certification system. The certification has created an entirely new legal structure for any company that wants to take their mission deeply into their financial DNA.[8] B Corp certification is not for any particular industry; it is for brands that care about doing good. Since its founding in 2007, the B Corp community has swollen to include over 3,000 companies in over 70 countries across 150 industries, from everything from clothing company, Patagonia, to local bakeries and art agencies.

Once companies earn a B Corp certification they can use the stamp to attract customers, future employees and investors. As a consumer, buying from any B Corp company should give you great confidence in that company's products and that you are voting well with your money.

If you work for a company that might be interested in becoming certified, or you have your own company, there are a few different steps of assessment which are available free online. Questions are broken down into governance, workers' rights, community, minorities, customer feedback, social and environmental impacts. Once the assessment is filled out, your company will be awarded a score out of 200 points – 80 or higher is necessary to be eligible. This might sound easy, but even companies that are already sustainability focused struggle when they take it for the first time. But don't be deterred; re-apply when you feel your company is closer to the requirements. Once the company meets the requirements, it is officially audited to ensure that all claims are accurate, and there will be a fee depending on the size of the company.

Taking the baseline assessment allows companies to be aware of where they can be doing better in business or ethics. For that reason, as an employee or business owner, it is in everyone's interest to aim high. Once your company is B Corp certified, you also get to be part of a community of like-minded companies to collaborate with. By joining B Corp, or at least striving towards it, you are benefiting not just shareholders but also all stakeholders involved. Your business will be changing the world, by conducting all business practices as if people and planet always matter and thus becoming responsible stewards for future generations. Bcorporation.net

The climate is changing – you can too

Reducing carbon emissions and changing systems, supply chains and global economies are macro level goals. On the everyday scale, each business could start with the smaller, more tangible, goals. If you have a product or service business, the real question is: does sustainability have an impact on brand value?

The answer is based on all the evidence of the dire state of our planet and an urgent call for action for a better world. A business will attract and hold customers by responding to this growing demand for sustainability, which in turn is linked to financial growth and stronger value. Whether you work for a large corporate, and could chat to the human resources department or your boss, or if you run your own business, here are some ideas to consider:

- Align your brand or company to a specific purpose – climate or other.
- Support green vendors and do business with green-friendly fair-trade companies.
- Promote a paperless office and only offer customers online receipts.
- Brighten up your office with green plants.
- Embrace renewable energy wherever possible.
- Minimise waste and implement a recycling programme.
- Have a no-plastic policy around the office and encourage everyone to bring their own reusable water bottle and coffee mug to work daily.
- Go carbon neutral.
- Donate a percentage of profits to a climate cause or environmental charity relevant to your company.
- Encourage volunteer days during working hours for employees.
- Encourage sustainable transportation by using public services or car-pooling with co-workers.
- Bring in inspirational speakers for your employees and start creating a more compassionate work culture.
- Align your businesses as much as possible with the United Nations Sustainable Development framework and Global Goals (see pages 33–34).
- If you manufacture items, consider producing them from

recycled products. Up-cycle or source materials locally wherever possible.

- Reduce plastic wrapping and use biodegradable packaging, such as envelopes made from sugar starch.
- Be more transparent with your supply chains and offer this information to customers.
- Ensure that every person in your supply chain receives a decent and fair living wage, no matter where they are in the world and what part of the business they intercept.
- Don't ever be afraid to weave your company's purpose into your brand's identity and everything you do.
- Get B-Corp certified.

The sustainability of business is a force for good; it is a moment of invitation for businesses old and new to examine the blind spots where they can change and be part of the solutions moving forward. Do you have a business idea? Do you have an idea of how to make something within your existing job and company more eco, ethical and conscious? No matter how big or how small your idea, jot down your green ideas in a notebook or at the back of this book.

7

People and Planet

'Climate change is not a fight for power, it's a fight for survival'

Oumarou Ibrahim

Looking towards a future where climate change and unprecedented storms are the global norm, the challenges will create a multitude of critical issues for people around the world, from large-scale human migration due to extreme weather events, to inter-state competition for vital resources such as food and water. Entire nations, in particular small islands, are disappearing under rising sea levels, and current legal systems are not equipped to support climate migrants. The impacts of climate change on people are only set to intensify in the coming decades and will drastically test the scope of global governance.

This chapter explores the 'people angle' of climate change, and also delves into gender and women's issues surrounding the topic. Gender determines what is expected, allowed and valued in a woman or a man in a given context. It determines responsibilities, opportunities, resources and powers associated with being a specific gender. Women are often in a disadvantaged position in many countries, particularly those in the developing world, so they have distinct vulnerabilities that shape the way they experience climate change and disasters. Yet the same destructive forces of climate disasters also create opportunities for women, as agents of change, to address gender disparities. Including the knowledge and voices of 50 per cent of our population in forming risk-reduction measures, as well as disaster response, is vital. The skill sets of women, including those who are the most affected by climate disasters, need to be capitalised upon. The signs of trouble further down the track as a result of climate change are becoming apparent, but the *human* disaster is happening now, in full view.

Women and nature

Ecofeminism examines the connections between women and nature. The term emerged in the mid-1970s and was first used by French feminist Françoise d'Eaubonne to bring together the two ideologies of feminism and the environment, as well as challenging both. It is based on a concern from the green movement about the impact of human activities on the planet, and, from feminism, it takes the view of humanity as gendered in ways that subordinate, exploit and oppress women.[1]

For more and more women across the globe, climate change

is a growing threat that disproportionately affects us. Whether in developed or developing countries, globally, women commonly face more poverty and have less socioeconomic power than men due to existing roles, responsibilities and cultural norms. In times of climate vulnerability, it makes it harder for women to recover from natural disasters that affect the infrastructure, homes and livelihoods. Women are more likely to make sacrifices to protect their families, such as eating less food and ensuring no one else is hungry. After a huge hurricane in Dominica in September 2017, 76 per cent of female farmers reported major losses, and some women couldn't immediately return to work because of damage to their homes as well. Most of the income women made from the farms was used to send their children to school. Until the women could grow their crops again it was impossible for them to get back on their feet for at least a couple of months.

Fast Facts:

- Disasters lower women's life expectancy more than men's, according to data from 141 countries affected by a natural disaster between 1981 and 2002.
- Women, boys and girls are 14 times more likely to die than men during a natural disaster.
- Following a natural disaster, it is more likely that women will be victims of domestic and sexual violence; many even avoid using shelters for fear of being sexually assaulted.[2]

Women can make a difference

At the beginning of the book I spoke about 'sustainable development', a key definition used by global governments and the United Nations to propel our world forward. If climate change policies are about ensuring a sustainable future by combining development and the environment issues, they are obligated to consider the interests of all stakeholders. This means placing women at the heart of all mediation, collaboration and visions for sustainable development equally, with men. Ecofeminists would argue that societies and systems will continue to favour men over women, *and* the planet, if women continue to be absent from leadership and decision-making positions and processes, and where the environment does not play a key role.

Fighting for equality is one thing, but increasing awareness of the institutional barriers that prevent the oppressed from reaching equality forces us to re-evaluate why society is structured in the manner in which it is and why we behave in the way that we do.

You may have read these first paragraphs and asked yourself, 'God, how has this book morphed into women's rights and "ecofeminism" – a serious and controversial topic, some might say?' As a female author I felt it my responsibility to dedicate a chapter to women and climate change, because it is frequently neglected in mainstream media.

By raising women's stories and experiences, I hope that you, as the reader, can begin to look at climate change through a new lens: looking at climate issues not only from what we see happening around us, but also from an underlying depth of feminism.

Added to this, we, as women, also have daily routines and actions that are distinct from men, and so our beauty regimes and lifestyles can have a positive impact on the environment.

This chapter aims to paint a picture of how climate change is affecting women globally and to bring it back home to the individual level and how women can be fiercely more sustainable in the choices they make for their health and in the home.

Climate change is a women's issue

I will never forget the day I read an article in the *Guardian* about the rise of child bride numbers across Africa because of climate change. The headline of the article seemed out of sync, but as I delved further into the research I realised that it was completely possible. Every year is getting hotter. Every year the rains are less predictable: they are either late and the crop fails or there is too much water and the floods sweep the fields away. Every year the yields are falling. Every year there are fewer fish in the rivers and the sea. Every year the families have less money, until eventually there seems only one answer: the girls must leave the family home as young as 13 and get married so that there will be enough food for the rest of the family to eat. In 2015 UNICEF warned that across Africa as a whole the total number of child brides could more than double to 310 million by 2050 if current climate trends continue.[3]

In central Africa, where up to 90 per cent of Lake Chad has

dried up, indigenous groups are at high risk. As the lake evaporates, women have to walk much further to collect water.

As temperatures across rural areas of Sudan have soared, uneven rainfall has influenced droughts, so the men of rural villages are migrating to cities in search of employment. The role of women farmers has, therefore, grown out of the impacts of climate change. More and more women can be seen digging up the earth and scattering seeds across the lands. Women, whose traditional roles have been caring for children and homes, have stepped into different roles for the family, such as becoming the providers of food, while their husbands are away trying to secure sustainable incomes.[4]

In the wake of the 2004 tsunami, the charity Oxfam reported that surviving men outnumbered women by almost 3:1 in Sri Lanka, Indonesia and India. Although there is no one particular reason for this, men were more likely to be able to swim, and women lost precious time when evacuating trying to search for their children and other relatives.

Destroying the patriarchy, not the planet

These stories are mentioned not to make you cry but to empower women to recognise our vulnerabilities within the climate space and to enable us to act for ourselves and each other. We women are half the world and yet there are huge disparities in representation within governments and organisations working on climate change.

More women's voices *need* to be included in policy, planning and decision-making processes at the highest level on the topic of climate change, in order to create new mandates for a brighter future. To begin seeing some of the facts and stories we have seen above diminish globally we need to ensure that the policies from the leaders we elect to represent us:

- Encourage female involvement – including indigenous women and grassroots groups – in matters of climate change negotiations and resource management. Provide women with access to resources and give them opportunities to participate in mitigation and adaptation processes.
- Ensure that key decision-makers understand how environmental degradation and climate change affect women differently from men.
- Promote gender-responsive approaches to climate financing.
- Invest in technologies and initiatives to enhance sustainable and renewable energy sources that reflect women's knowledge, needs and roles.
- Integrate environmental conversation strategies within family planning and women's health programming and vice versa.[5]

Sabita – safety coordinator

When we apply the above by including and providing women with tools to act in the face of climate change, we witness stories like this one of Sabita from Bangladesh.

> Sabita, 42, is no ordinary mother of three. When a cyclone struck her village last year, her forethought, in running emergency cyclone drills in preparation for such storms, helped to save the lives of 500 people. She lives in one of the most remote parts of southern Bangladesh, which is increasingly experiencing dangerous weather driven by climate change. ActionAid trained her in setting up a network of women who alerted each other when the storm was coming and calmly got people to safety when it hit. Sabita and the other women have made a pact that they will team up again in the face of any other climate emergency.[6]

Eco-warriors

There are women around the world working tirelessly to create a better planet, and this chapter wouldn't be complete without mentioning a few of those inspiring change-makers. These are women in the environmental space who are forging new paths of creativity, education, science and activism and pushing every day to protect our Earth and future generations. Some of them we can thank for opening up opportunities and we can reinforce the groundwork that they have put in; others we get to watch and see them grow as the environmental and conservation movements strengthen. Here are some women warriors at the top of their games revolutionising our world:

Jane Goodall, UK

In 1960 Jane Goodall set out to study wild chimpanzees. She immersed herself in their lives, bypassing rigid procedures to make discoveries about primate behaviour. She has continued to shape scientific discourse as an author, filmmaker and advocate for ecological preservation through the Jane Goodall Institute, an NGO that focuses on the education of youth, conservation and strong environmental practices around the world.

Greta Thunberg, Sweden

Greta Thunberg has become one of the world's foremost environmental activists through her weekly Friday for Future protests. At 16 years old, Greta was nominated for the Nobel Peace Prize and she has spoken at the United Nations, the World Economic Forum and in front of governments globally. Best known for speaking boldly on what she calls the 'climate crisis' and holding politicians to account for their lack of action, Greta now has millions of young people supporting her movement.

Dr Sylvia Earle, USA

Sylvia Earle has been at the forefront of ocean exploration for more than four decades. From her beginnings as the captain of the first all-female team to live underwater in 1970, Earle has led more than 50 expeditions and clocked more than 7,000 hours underwater. In 1979, she walked untethered on the sea floor at a lower depth than any other woman before or since. In the 1980s, she started the companies Deep Ocean Engineering and Deep Ocean Technologies to design undersea vehicles that allow scientists to work at previously inaccessible depths. In the early 1990s, she served as chief scientist of the National Oceanographic and Atmospheric Administration. Dr Earle has recently used money from a TED prize to fund a non-profit called Mission Blue, dedicated to creating protected marine reserves around the world.

Isabel Wijsen and Melati Wijsen, Indonesia

Isabel and Melati started Bye Bye Plastic Bags at the ages of 10 and 12, after being inspired by a lesson in school about significant people such as Nelson Mandela and Mahatma Gandhi. The sisters returned home that day and asked themselves, 'What can we do living in Bali, *now*, to make a difference?' They were frustrated by the negative impact of plastic around them. Indonesia is the second largest plastic polluter in the world after China, and their local government declared a 'garbage emergency'. Their goal was to convince Balinese residents to say no to plastic bags, and thus Bye Bye Plastic Bags, a youth-led NGO, was born and has now grown into a well-known international movement of inspiration and youth empowerment. Their movement led the way for an official ban on single-use plastic bags, straws and polystyrene across the islands of Bali.

Christiana Figueres, Costa Rica

'Impossible isn't a fact, it's an attitude', so says Christiana Figueres, the woman tasked with saving the world from global warming, who has helped the world to change its attitude on climate change. Christiana was in charge of the pivotal Paris Climate Agreement and made history by producing a landmark deal to limit future carbon emissions. She was appointed Executive Secretary of the UN Framework Convention on Climate Change and delivered, for six consecutive years, global negotiation sessions, convincing governments that it was in their national interest to engage in sustainable development. The results of her actions? One hundred and eighty nine governments submitted comprehensive climate change plans, some of which are legally binding, to the UN. She is a powerhouse of optimism and hope, a champion of young people and clean technologies, and a woman determined to leave the world in a better state than it is now.

There is no 'other'

On the positive side, there are areas in our daily lives that differ from a man's, such as beauty, babies and menstruation, where we, as women, can empower ourselves to reduce our personal waste and become more sustainable. When we reference eco-feminism, it's all about inclusive thinking from both sexes, and therefore conversations about babies, periods and standards of beauty require a special strength and integrity from both sexes in understanding the complexity of these issues.

The actions and decisions of past generations have not made Earth a very nice place for us, and with so much uncertainty ahead of us all, some people known as 'Birth Strikers' have decided not to have children in response to the coming climate breakdown. This is a radical acknowledgement of how the existential threat is already altering the way some of us imagine our future. Birth Strikers hope to channel their grief into something positive and hopeful, by creating a political movement to raise awareness of the severity of our climate crisis, as well as becoming a support group for those who have decided not to bring children into the world.

#BirthStrike

Declaration

We, the undersigned, declare our decision not to bear children due to the severity of the ecological crisis and the current inaction of governing forces in the face of this existential threat. Insecurity of future, despair at our species' relationship to our habitat and each other, channelling time into activism and rebellion, are all common motivations.

What do birth strikers do?

BirthStrike shows how the threat of ecological disaster is altering the way we imagine our future. It is a radical acknowledgement that our planet has entered a Sixth Mass Extinction

> event due to man-made impacts on the environment. BirthStrike stands in solidarity with the environmental justice movement, the academic and scientific community who are encouraging acts of rebellion and widespread system change in order to urgently save our future.[7]

To many, it might seem radical to not have children in favour of the environment, but top thinkers such as Sir David Attenborough have highlighted population growth as one of the critical threats to the future of our planet. With population skyrocketing, babies born every day mean less food for other people, with more pressure on natural resources and potentially sentencing those children to a life of uncertainty on an unstable planet.

We should not be in a situation where some of us feel genuinely scared to bring life into the world. For some women, fears about our environment are so strong that they have decided not to have biological children. For many it has been a painful emotional choice not to give life, and there has been a significant violent backlash online against women speaking out about their choices, but there is now a growing community of outspoken activists sharing their experiences.

Would you join the #BirthStrike?

At the opposite end of the spectrum, looking into ways of family planning, informing women and girls globally and in the developing world about reproductive health can help diminish the strain on the environment. Data shows that the higher the level of a woman's education attainment, the fewer children she is likely to bear. Women who are educated post primary school and who have the ability to read, tend to learn different ideas of desired family size through school, community and exposure to global communication networks. When women are educated about contraception, they can intentionally plan their families and provide a stronger framework for future generations.

Periods and the planet

Unmentionable? Absolutely not! We need an urgent conversation about periods.

The average woman, transgender or non-binary person who menstruates, will spend an average of 2,280 days on their period and use more than 11,000 tampons or pads over the course of their lifetime.[8] Menstruation is a natural and healthy aspect of a woman's life; however, while in many cultures periods are celebrated, in many others the topic is still taboo, and period shaming is having a huge impact on women and girls globally. As a result, menstrual products and our knowledge of how to use different products is affecting our health and environment.

The plastic side of menstruation

Do you know what a regular pad or tampon is made of? Would you be shocked to learn that menstruation products contain plastic?

The average sanitary pad contains the equivalent of about four plastic bags.

Many of us are completely oblivious to the risky ingredients that sanitary products can contain and what happens to them once we throw them away. In fact, manufacturers are not required to disclose the ingredients in tampons and pads, but what health experts and sustainability advocates have discovered has led them to raise the alarm.

Generally, tampons and pads are made of non-organic cotton that contains toxic non-biodegradable plastic, synthetic materials such as petrochemical additives, glue, bleach, wood pulp and paint stripper. Excuse me – what?

Some of the pesticides used to grow the non-organic cotton are even classified by environmental agencies around the world as carcinogenic, and, according to the World Health Organization, some of the toxins found have been linked to reproductive infertility problems, cancer and the increased risks of disease. So, yeah, now you know what exactly is looming in your average supermarket pad and tampon – it is time to switch. *Now.*

The ecological low-down on sanitary products

That was only the health part to grab your attention, and if you're still able to read on, let's talk about periods from an ecology perspective. The amount of pollution pads and tampons generate yearly is cause for concern, especially when their compositions are non-biodegradable.

The average person who menstruates throws away up to 200kg of menstrual products in their lifetime.[9]

It is estimated that globally over 100 billion menstrual hygiene products are disposed of annually. Most are not recyclable, either, because they have been in contact with our blood, so therefore they simply end up in landfill. Because the products contain plastic, it will take over 500 years for each pad and tampon to break down. When incinerated, as is the case for most landfill around the world, the products, of course, release carbon dioxide among other toxic fumes. Menstrual hygiene products generate a total carbon footprint of about 15 million tons of greenhouse gas emissions annually, according to the United Nations Environment Programme. That is the equivalent of burning up to 35 million barrels of oil.

Although the fight against single-use plastics, such as bags and straws, has become a mainstream issue, the conversation around menstrual products is a little trickier. Talking openly about menstrual waste in most cultures is difficult. I mean, let's be real: in the Western world up until recently, adverts that focus on product discretion on TV have been depicting blood as a mysterious blue liquid. As a society we avoid talking real about periods, thus menstrual products are incorrectly disposed of because women are simply not educated enough from a young age. And, don't forget, from earlier chapters the plastic in these products breaks down into microplastics that linger in our waterways for decades and potentially end up in our food systems. Menstrual products are always found in the top five items at beach clean-ups.

Contributing to the plastic pollution problem is just one factor, but menstrual products also pose the threat of spreading diseases and pathogens because of the bodily waste that they contain, putting communities living nearby these waterways at high risk. But how do they even get in the waterways in the first place? A lot of people flush tampons and pads down the loo; no guilt-tripping intended here, but, please ladies, don't flush your products but dispose of them correctly.

It's time for women to take our monthly menstruation into our own power, to be in charge of knowing what we put so close to our intimate parts. We need to know that we, as women, can be part of the environmental battle by choosing – and disposing of – our monthly products wisely:

- Never flush your menstrual products down the toilet. *Ever.* Please dispose of them responsibly in a waste bin.
- Consider reusable menstrual hygiene products such as menstrual cups, period underwear and reusable sanitary napkins. These all sound a little uncomfortable, but go on and try. It might also take time to test what is better for you, so don't lose hope if the first time doesn't work out. These reusable items will also save you money in the long run.
- When disposable brands are necessary, choose organic and transparent brands such as these ones that are better for your health and the planet: OHNE, TOTM (Time of the Month), DAME, Freda, Luna Pads.
- Help to make menstruation a bigger public issue by:

1. Demanding our governments force corporates to disclose what ingredients are in our menstrual products.
2. Urging companies *not* to use plastic in their products.
3. Lobbying governments for free sanitary products for women and girls globally to help to put an end to period poverty.
4. Support and check out City to Sea, a group fighting ocean pollution that started a #PlasticFreePeriod campaign.

Beauty

The beauty industry creates large quantities of waste for disposal. One hundred and twenty billion units of packaging are produced every year by the global cosmetics industry,[10] most of which is not recyclable and takes hundreds of years to decompose. Too many products come wrapped in plastic, stuffed with paper inserts, cardboard sleeves, mirrored glass, and sometimes all of the above in one single purchase. Of the products we can recycle, many of us don't because we simply don't remember to recycle our bathroom goods. Or, for those items that do make it to a recycling facility, the industry often doesn't have the patience to recycle mixed-component materials.

There are, however, many planet-friendly beauty options available today, from high end to high-street to niche brands. Take a conscious decision to shop your beauty items responsibly with less packaging and more plant-based ingredients. As long as the product does its job on our faces and body, who cares if the packaging looks ugly if the future of beauty looks green?

- Top up, don't shop. Many companies are offering ways to refill your products instead of replacing them, either over the counter or through 'circular' shopping platforms including home-delivery services.
- Recycle and refill. If buying refillable bottles in the first place is not possible, you will find that navigating beauty recycling just got easier, because brands are starting to provide in-store bins and drop-off points for empties.
- Be picky with your plastic. OK, throughout this book it's all about optimistic swaps, but the tips are also trying to be realistic. Fully sustainable products might not be the most convenient for you and there might be slim pickings in your shopping sphere, so at least look for the recycle symbol.
- Get creative with how you can reuse your packaging.
- Buy lip balm in a tin container rather than a plastic tube.
- When it comes to removing make-up, instead of using facial wipes and disposable pads, opt for reusable cotton pads or sponges made from vegetable roots instead of synthetic fibres.
- Stop using single-use wipes and baby wipes. Once upon a time we thought that cosmetic wipes deserved a Nobel Prize for services to busy women. But now? Now we know even the so-called biodegradable sheets end their days in landfills, or most annoyingly causing 93 per cent of sewer

blockages in the UK.[11] Stick to hand-wash muslin cloths, which are also dermatologist recommended as kinder to our skin.

- Sparkle with care. Glitter might scream *festival fun!* but it's a no-go if you're going green. Regular glitter should be a no-no, because it is made up of tiny plastic particles. When we wash it off it makes its way into our waterways and oceans and is swallowed by wildlife. Still want to sparkle? Opt for biodegradable alternatives, such as Disco Dust, Eco Glitter Fun or Bleach London's body glitters. There are plenty of alternatives to find, so don't hold back next festival season.
- While you're taking a shower, switch out your usual bottled soap and shampoo for hard bars instead.

SHOWER SAVINGS

Beyond the packaging, the average Brit gets through 142 litres of water per day, using an average of 50 litres of water per shower, which simply goes down the plughole. The brand Aveda worked out that we could save approximately 27 litres of water by reducing our shower time by just three minutes and going an extra day without washing our hair. While you're saving resources, give a thought to the 840-plus million people who live without access to clean water.

Spraying pollution

Another negative impact of the beauty industry is the air pollution created by the likes of perfumes, hairsprays and spray deodorants. Studies have highlighted that scented products emit the same level of chemical vapours as petroleum emissions from cars. The particles in the air are harmful for the ozone layer and our lungs.[12] You can make a difference:

- Switch to roll-on deodorants and perfumes.
- Choose oils and gels as alternative hair-care products to sprays.

Harmful sunscreens

The beauty industry came under fire in recent years for using problematic ingredients. Oxybenzone, a chemical often found in sun creams, is extremely harmful to coral reefs and ocean ecosystems. About 14,000 tonnes of chemical-filled sun cream washes off our skin and into the sea each year, making our sun-protection habits a major component of ocean damage. What you can do:

- Buy brands that have banned the most corrosive chemical filters from their formulas while keeping the SPF factor high and the formula easy to apply.
- Buying natural products from smaller companies means that the ingredients are more likely to have been sourced

> sustainably, as supply chains will be shorter and more transparent for you to go and check. Palm oil, for example, is used in approximately half of all consumer goods. It leads to widespread deforestation and even the extinction of animal species. To avoid hidden problems of this kind, the best way to identify if a product's ingredients are sustainably sourced is to look out for the Fair Trade and Rainforest Alliance logos on the packaging.

By reading this chapter you have become an eco-feminist without even realising it. When you practise eco-friendly habits and identify with a healthier planet, you are contributing to the betterment of *all* human beings and other creatures.

> 'We are just 1.5 metres above sea level and anything over that, any rise in sea level – anything near that – would basically wipe off the Maldives . . . We have a right to live.'
>
> **Mohamed Nasheed,**
> **former president of the Maldives**

People are affected everywhere

Climate change is indifferent and it affects *everyone*, no matter what side of politics you are on, what faith you hold or even whether you believe in democracy. However, we are living in a world where it is easy to blame someone else:

environmentalism is another person's problem and that 'other' person will also solve it.

For those of us living in a big city, the idea of someone's home being washed away during a heavy and unexpected rainstorm, or an entire island nation disappearing by 2050 because of rising sea levels, seems completely unimaginable. We are already seeing natural disasters, famine and floods as a result of climate change, and more will come in the future, pushing more countries into wars and conflicts, and creating disillusionment over the modern-day political narrative. Mozambique, Zimbabwe and Malawi were devastated by cyclone Idai in 2019. In Syria, Libya and Yemen, climate chaos has partly contributed to civil war. In Guatemala, Honduras and El Salvador, crop failure, drought and the collapse of fisheries have driven people from their homes – despair is not an option. The ability of countries like those mentioned to bounce back is difficult due to their unstable political conditions and underlying corruption, but the Earth's climate will ultimately command more political attention and stronger international relations.

The common theme throughout this book has been to empower each of us to take individual action, in recognition that small changes multiplied by thousands are impactful. Our inaction has forced people across the world into action, as they respond to terrifying circumstances caused primarily by our consumption in the West.[13] By making eco-switches we start taking accountability not just for ourselves and our own social circles but for every living thing on the planet. It's hard for us to see the damage happening right here right now in our personal circumstances, but if we can channel empathy towards those people facing climate injustices, we can help to build a real network of change.

As the author of this book, I have consistently emphasised that I come from a place of privilege and, fortunately, living in London, I do not yet have to worry about migrating somewhere else in the coming years and have not yet experienced a natural disaster. Through travelling and working in this space, I have witnessed the impacts of climate change on people and communities, but this is not my story to tell. Here are the voices and stories of people around the world in their own words.

Selina Neirok Leem, climate warrior from the Marshall Islands

'We have eleven years left until we surpass the 1.5°C.' This statement by a Norwegian scientist echoed in the hollows of my head as I stopped listening to the rest of the event. I looked at the faces, hearing but not listening.

My home. My beautiful, precious home. Mama, baba, bubu, jimma, relatives, friends, neighbours, acquaintances – all these faces that I am fighting for. Have we lost? Is this it?

The Marshall Islands, an island nation located halfway between Hawaii and Australia, is home to approximately 52,000 people. My home faces droughts, soil and water salination from seas washed inland, sea inundations and erosion. We, the occupants, face waves higher than our sea walls, broken houses, graves pummelled with a force only the sea is capable of, unearthing our family members' remains, and dried and salinated crops, and underground water. Again, the next day, after a rough night of land-and-water tousling, we pick up the remnants. We build the sea wall that was broken, clean the trash the water had vomited back to us, mop and sweep the water that had gotten into the house. We pray for our well-being together, and privately. Trying to mask our fear from the others.

Home faces extinction and we say no to that. Our leaders and youth leaders have become climate warriors across the world and at home as well. Across the seas, we speak to thousands of deaf ears, demanding climate justice for our home. At home, we educate and nurture our youths on climate change, the tools that they have and can use to mobilise themselves to help us adapt and mitigate damages. Home is resilient, and so shall we be. We will not give up. You never gave us the choice anyways. We do not have that luxury. We are doing what we can. It is past time that you should do your part as well. This is not an ask – this is a demand.

Farman, Chitral, Pakistan

I am Farman. When I was a child we would get snow in the winter, light rain in spring or autumn and no rain in the summer. But now, what is happening is in the summer we are getting sudden and heavy rainfall causing flash flooding. In the last five years, the weather has changed a lot.

We are experiencing flooding, prolonged heatwaves, sea- and river-level rise, excessive rainfall, and hurricanes, and, owing to temperature increases, our glaciers are melting. There is a total of 282 glaciers in the region – 187 of them have already melted and another 20 are declared dangerous. The melting glaciers lead to flooding, which washes away everything in its path.

Due to the heat, we don't get snow in Chitral anymore; however, in the very remote villages higher in the mountains they get heavy snow – and, suddenly, without warning. They also get heavy rainfall, which causes avalanches. In March this year, nine students died in an avalanche. They had just finished their exams and were walking home.

In the cities, the heat is causing other issues. Poor people lack electricity, so there is no cold water. The water that comes out of our taps is very hot, and there are issues with water sanitation so it's not safe to drink. People are dying from heatstroke, and this is increasing every year.

Everything is affecting the poor in every way, as they don't have basic facilities. Children are dying from diarrhoea due to water contamination from flash flooding, and there is a whole range of other waterborne diseases affecting the local people.

Pakistan is one of the top ten countries hit hardest by extreme weather events and it ranks eighth on the Global Climate Risk Index. It is estimated that there are 487 deaths annually linked to extreme weather-related incidents. There were two heavy flash floods in my village in 2015. We were cut off from neighbouring towns and villages for 2 months as

> roads and bridges were washed away. We have not had any electricity for 12 months, because the entire hydroelectric plant was washed away. Now we use wood for our stove and I have a small solar panel, which allows me to power three or four lightbulbs in my house.
>
> I am feeling good that I am sharing my story.[14]

Disappearing communities

Can you imagine worrying whether in 30 years' time you will have a home, and even a country in which your own children can grow up?

For many small-island developing nations, such as the Marshall Islands where Selina comes from, sea levels are rising so rapidly that islands like hers might no longer exist by 2050 or soon after. Entire nations are disappearing. If their passports are no longer valid because their country sank under the sea, where do they go? Where do they belong to? There are no current decrees under international law that protects climate refugees, so who from the international stage is going to step up and help? We are already living in a time of so much political turmoil and tragedy, how much more can nations handle? We need to put measures in place, a plan to safeguard future generations.

For small islands, climate change presents unique challenges. The difficulties that all countries face in effectively coping with climate change impacts are exacerbated because of their small geographical area, isolation and exposure to the natural elements, most prominently increased temperatures, more frequent and extreme weather events and rising sea levels. All these increase disease risk, illness and deaths. Adapting to ongoing and future climate change is critically important for all small islands, and a huge thank-you goes to the next generation of leaders like Selina for sharing her story of what the reality is like.

Climate refugees

According to the Environmental Justice Foundation, climate refugees are persons, or groups of persons, who, for reasons of sudden or progressive climate-related change in the environment that adversely affects their lives or living conditions, are obliged to leave their habitual homes either temporarily or permanently, and who move either within their country or abroad.[15]

The idea of climate refugees is not too far away from home either. One and a half million homes are at risk by 2080 in the UK alone. Fairbourne is an idyllic Welsh village nestled on the Irish Sea coast, bordered by water and mountains. But after a sea wall was breached in 2014, their County Council decided it would no longer repair the defences indefinitely and, given the anticipated rise in sea levels generally around the world, they announced that Fairbourne would be 'decommissioned' starting in 2045. What does this mean? The current 1,200 residents of Fairbourne will be forced to move out and, as of today, officially labelled the UK's first climate change refugees. As Fairbourne prepares to remove all trace of human existence and the land eventually returns to the sea, these won't be the last UK residents to meet this fate, as many other villages lie along England's eroding coastlines and flood plains.

The reality is, though, climate change does disproportionately affect poorer regions of the world, frequently hitting those communities that contributed least to climate change. The magnitude of the urgency means we need to realise that climate change is a problem for everyone, at both ends of the wealth spectrum. With that in mind, we, as individuals, have a moral responsibility to give a shit.

How can we be so technologically advanced and progressive, yet be in a situation where human rights in some countries are retracting and people are starving when we *do* have enough food to feed the world? Children are dying of preventable diseases, and now more freak weather patterns are hitting communities and destroying absolutely everything. Through the rise of capitalism and consumerism, we are shifting into a zero-sum game model of existence, a paradigm in which one person's gain is equivalent to another's loss, so the net overall change in any benefit is, in fact, zero.[16] You could think of this in terms of a person in a developed country using more energy and causing sea levels to rise on a small island nation and another person's home literally washing away. Or it might be one person chopping down part of a forest to build a home, resulting in the death of plant and animal species.

This chapter might sound dramatic and apocalyptic to some readers. You might live in a city where something like this will probably never touch you, or wonder why you should even feel emotionally attached to such things happening. We might relieve ourselves of responsibility by not seeing the visible effects where we live, and say that it's too late to act, but in doing so we condemn others to increased poverty, and even death. The circumstances brought about by extreme weather are beyond our control, but those before us caused the irregularities, and now it's time to reverse those actions.

The climate is changing – you can too

By acting on *any* of the tips in this book you will be showing your compassion for the environment, and by helping the planet you are helping people globally, but you can take it further and be part of people power:

Read outside your normal news bubble and check out what is happening around the world.

Call on governments to recognise climate refugees and support a new legal agreement to guarantee their rights. These populations currently have no support under existing legislation, and organisations such as Environmental Justice Foundation are doing incredible work to help get their voices across: www.ejfoundation.org

When you see a story about communities globally affected by climate change and natural disasters, share the news with your own circle or across your social channels. When the Amazon started burning and the mainstream media wasn't covering the destruction, people took to social media, creating a huge uproar. By highlighting areas that might need specific attention when global media are not doing their job, you are being a person who cares.

Watch more documentaries and mini-series on the impacts of climate change.

Follow movements such as Extinction Rebellion and Greenpeace to increase collective people power and feel part of something bigger where people, together, are fighting for climate change.

> 'Sometimes when you want to make a change, then it is necessary to turn the world upside down because it is just not for the better, but it is simply for the best'
>
> **Selina Neirok, a small-island girl with big dreams**

Conclusion: The Climate is Changing – Are You?

'Whatever you do (may be) insignificant, but it is very important you do it'

Mahatma Gandhi

We live in an incredibly complex world where no one answer across any industry can immediately solve the climate crisis we are currently facing.

Being meaningful is to be part of the solution.

We cannot achieve sustainability while we ignore inequality.

We have to fight ignorance and share knowledge and resources, and pave the way for those behind us.

A warning to humanity: We have just 12 years to turn things around before we pass the point of no return.

The Earth is our mother and we must take care of her.

Be a good human.

Be happy with your choices.

Share.

Wanting a better planet starts with choosing differently.

Refuse. Reduce. Recycle.

Always remember that small acts multiplied by millions of people can change the world.

Your own optimism is what will support the real infrastructure of global change.

Time to put the negativity aside. Devastating news, science reports: use them as catalysts for change.

Socially responsible behaviours are en vogue.

Reconnect with and rethink what you already have.

Sustainable development is the kind of development that satisfies the needs of the present without adversely affecting the ability of future generations to satisfy their needs.

Never ever forget the people aspect of climate change, because it could be you, your family or community that is next in the firing line.

'This is the moral challenge of our generation. Not only are the eyes of the world upon us. More important, succeeding generations depend on us. We cannot rob our children of their future.'

Ban Ki-Moon, former UN secretary general

The starting point

Research into complex systems reveals that our planet is on the edge of ecological devastation and a brutal surge in inequality. Scientists at the highest level have issued the warning that our planet is facing a dire climate emergency, and we have a warning to humanity that we have just 12 years to turn things around before we pass the point of no return. The topics touched upon in this book merely skim the surface of what is happening throughout different industries and their contributions to climate change, giving a global picture of current and future potential hope and alternatives. If we continue the way we are, producing as many clothes every year, creating as much food waste and discarding single-use plastics the way we do, our ongoing destruction of nature, forests, oceans and the capacity to breathe safe air and drink clean water will be in jeopardy. We are taking without giving much back.

The goal has been clear from a science, business and policy level for over 30 years, and none of the above have essentially made any progress into protecting the world. When it comes to voting with our money, brands that pass the costs to consumers are a cop-out. Being mindful, where possible, of where and how you shop is incredibly powerful, as you are contributing to a sustainable consumer society. Sustainability is not a phase or a trend; the future of business entails having purpose, ethics and environmental considerations at the core.

We all have to eat, we all have to get dressed, we all need energy and we want toxic-free homes for our families. For the industries referenced in the book, from fashion to food and plastics, sustainability presents new business opportunities. Whereas there are greater emphases and incentives in the above sectors, the pace is too slow, and so now the responsibility to act falls into your hands.

Lots of governments, international organisations and businesses continue to talk, negotiate and even host events around the world with influential people debating climate change, promising action and a brighter future. The UN reports talk about the ruinous consequences of the 2°C and 3°C of global warming and planetary ecosystems coming to breaking point. The truth is, it's already happening, and we are seeing the effects in different parts of the world when we listen to the stories of communities losing their entire livelihoods because of pollution and changes in weather.

Today, people's homes are literally washing away; natural disasters are sweeping over regions at an unprecedented rate, resulting in millions of refugees worldwide displaced because of climate change with not enough protection under international law. Thousands of young people from small-island developing nations are petrified about where they will belong in the future. Climate change is harming the most vulnerable people globally, but never underestimate that you, or your family or community, could be next in the firing line of unpredictable weather and a shifting environment.

In the opening chapter, I explained that this book is a voice for the people whose lives are dramatically changing because of climate change. The areas first and most affected are the poorest regions of the Earth and the least represented on our planet who don't have the platform to share their stories.

This book is for those people that aren't even here yet; they haven't been born yet. It asks you to feel empathy for the unborn and the poor of the global south and change your lifestyle on the basis of their needs – something that humanity has never done before. The time to start caring for more than our immediate self, starts now.

The curve balls of history are everywhere, and no one can accurately predict when the next civilisation will come undone. The climate saviour will be a nonlinear transformation and will happen as the result of the gritty and persistent actions of a critical mass of engaged individuals like you. Regardless of seemingly overwhelming odds, there is no history available or road map for those doing the climate talking; there are no books or ancient documents explaining what to do when the planet is at tipping point. Reading this book, you might have gone through an array of emotions, fuelling consumption-guilt, sadness, depression or confusion at the actual state of our planet. The intention was never to be on the sorry side, though; the intention of this book is to create a tool of empowerment and knowledge, and a guide for you to make the world a better place through your choices.

Climate change is not a class affair, either; it is a matter of life or death for *all* life on our planet. Where do we start? By giving a shit, by having empathy, by supporting conscious companies, and by using and having what is essential now without compromising the needs of future generations. Empathy, compassion and humanity – three words that, if implemented, will be the foundation of the environmental revolution that we need so very much.

It's not about becoming so overly conscious that we put ourselves in the difficult position of dealing with a problem that is too big for our brains to process; that way lies cognitive overload that obstructs climate action. It's also not about thinking that climate change is someone else's responsibility and thus a reason not to act. Many of us in the world are comfortable; we feel that our own ecosystems are stable and secure because the effects of climate change are not bang in our faces. Due to this security we can be resistant to change, but when we start to link climate change effects within systems that are close to us, we can start to endure the restructuring of our lives.

Sometimes, history has a way of forcing ordinary people to face up to a moral encounter with a fate that they never expected. Do you want your grandchildren to turn around to you one day with abhorrence, asking, 'Why did you do nothing when there was still a chance to stop the horror?' We need to be prepared to face future generations when they ask what we did when we knew our future was on the line. If you have money, social and cultural capital and housing stability, the chances are you are going to be more likely to be in a position of privilege to adopt more of the switches advised in this book.

We are significantly interconnected more now than ever before – our lives are so intersectional. Empowered by unprecedented access to tools and resources, with a click of a finger we can learn about almost anything. Thanks to globalisation and the Internet, we are the most informed generation. Our world is so technologically connected that we are borderless. We are not neatly confined behind boundaries – every act we make, everything we say, is felt globally. For most of us we have so much autonomy and control over our own lives. For those of us lucky enough to live in democracies, our voices are big and loud. We can use tools such as social media, not only to connect but to call out injustices, call out brands, demand more from leaders, brands and peers about what we do, and what we don't, want in this world.

This book is written to help break down stereotypes. My story began as one woman who was able to change, from the inside out, an industry by using innovative means to raise awareness and open the discussion about climate change. I'm not a climate activist making money or a digital career out of advising. This book came from a place of not fitting in, still caring, and acting wherever and whenever possible. Caring for people and planet is about sharing ideas. It's about understanding different perspectives. It's about creating dialogue and bringing together even the most disparate groups. It's about recognising that at a global scale climate change is the greatest threat to our generation – recognising that *you* are the change.

We are too often told to be a certain way and be more committed to audacious and ambitious actions. There has been an atmosphere for so long in climate activism that there are only two camps in this argument: it's about 'us', as in climate believers, versus 'you', the everyday citizen – the 'good' versus the 'bad'. We are often told it's the government's job to solve the crisis and if you do care about the environment the only way is radical individual action. Become zero waste, vegan and as off-the-grid as physically possible or face the catastrophic consequences of a dystopian climate future. We do need radical action, yes, but each individual needs their own catalyst and starting point to their environmental journey. I hope this book has encouraged you to start somewhere or progress your existing journey – by doing that *you* are part of the climate solution to protect the planet for future generations.

The cost of inaction is too great. It's important to remember the severity of the situation, even as you sit comfortably reading this book. If every consumer immediately stopped using cars and planes, buying fast fashion and eating meat, and ditched plastic, the problems would not resolve overnight, because these efforts are but a drop in the ocean of the magnitude of damage that humans have caused our planet – and in such a short time frame. But what we *can* do that will be just as powerful is to ensure that we foster an environment that endorses these changes and educates the generations to come. We can now build the infrastructure for humanity to move forward, rewrite our wrongs, and invest in ideas and solutions that can build a brighter future for our children.

Enough hatred, enough judgement – the time to act is now. You on your own journey with what is right for you. No matter who you are, or where you are in life, and how much money you have, you can do your part to help save the planet. There are way more tips and tricks to live a more sustainable life than could cover all the pages available. Whether you make one change, five or ten from this book, I hope it leads you to feel that you've reached a higher purpose.

Are you changing?

Hope is sitting back. Courage is getting up and acting.

If a specific topic covered in this book really stood out for you, start your own campaign, whether it's a no-plastic policy in the workplace, or encouraging greater ethical fashion choices, or simply improving your diet by switching to buying locally and more plant-based food. Start today by asking yourself: if things could change, how and what would your ideal outcome be? Your voice and actions are the most powerful things you have.

- Start discussing your thoughts with friends and family.
- For a greater reach, use social media to ask questions and gauge opinions and support.
- Write to your local MP and the government.
- Write to your favourite brands.
- Call them out on social media.
- Join a protest.
- Mobilise together.
- Defend together.
- Educate together.
- Act together.
- We have each other, and that makes us together.

Remember: this a journey. Making significant lifestyle changes, however big or small, won't happen immediately, and so, as you begin your process, here is some space to take some notes about the following:

Establish *why* you care about the environment, people and future generations.
Assess your lifestyle, habits, spending and waste patterns.
After all, the climate is changing – are you?

Climate Change Thoughts

Further Reading and Resources

Books

This is Not a Drill by Extinction Rebellion

No One is Too Small to Make a Difference by Greta Thunberg

The Uninhabitable Earth: A Story of the Future by David Wallace-Wells

21 Lessons for the 21st Century by Yuval Noah Harari

Climate Justice: A Man-Made Problem With a Feminist Solution by Mary Robinson

This Changes Everything: Capitalism vs. the Climate by Naomi Klein

The Deep: The Hidden Wonders of Our Oceans and How We Can Protect Them by Alex Rogers

How To Make a Difference by Kate and Ella Robertson

Podcasts

'Talking Tastebuds' by Venetia Falconer

'Wardrobe Crisis' by Clare Press

'The Disclosure Podcast' by Earthling Ed

'Deliciously Ella' by Deliciously Ella

'Today in Focus' by the *Guardian*

'Getting Curious' by Jonathan Van Ness

Social media accounts

@trashisfortossers

@chicksforclimate

@futureearth

@oceangeneration

@muthaearth

@earthlinged

@jackharries & @finnharries

@sdgaction

@unep

@oneyoungworld

@mrspress

@fash_rev

@ecoage

@carly_bergman

@savingthegrace

@allhandsandhearts – if you are ever considering an opportunity to volunteer overseas, the charity organisation All Hands and Hearts serves the immediate and long-term needs of communities impacted by climate natural disasters.

References

Introduction

1 www.theguardian.com/environment/2018/oct/30/humanity-wiped-out-animals-since-1970-major-report-finds

2 www.vox.com/energy-and-environment/2019/5/17/18626825/alexandria-ocasio-cortez-greta-thunberg-climate-change

Chapter 1

1 www.bbc.co.uk/news/science-environment-11833685

2 www.bbc.co.uk/news/science-environment-11833685

3 www.wrap.org.uk/about-us/about/wrap-and-circular-economy

4 dictionary.cambridge.org/dictionary/english/deforestation

5 skepticalscience.com/climate-change-global-warming-basic.html

6 www.bbc.co.uk/news/science-environment-11833685

7 www.cencoos.org/learn/oa/intro

8 'Our Common Future Report', Brundtland Rep

www.sustainabledevelopment2015.org/AdvocacyToolkit/index.php/earth-summit-history/historical-documents/92-our-common-future

9 www.sciencealert.com/here-s-how-biodiversity-experts-recognise-that-we-re-midst-a-mass-extinction; https://www.nationalgeographic.com/science/prehistoric-world/mass-extinction/

10 www.nationalgeographic.com/magazine/2018/07/embark-essay-climate-change-pollution-revkin/

11 www.nrdc.org/stories/paris-climate-agreement-everything-you-need-know

12 www.bloomberg.com/news/articles/2018-09-06/great-barrier-reef-showing-signs-of-recovery

13 www.un.org/sustainabledevelopment/climate-negotiations-timeline/

14 www.businessinsider.com/ap-timeline-of-key-events-in-un-effort-against-climate-change-2015-11?r=US&IR=T

15 www.ajc.com/news/world/what-the-paris-climate-agreement-things-you-should-know/zG6aSruOocgyOvTd4VnPTM/

16 BBC Online article: 'Final Call to Save the World From Climate Catastrophe' www.bbc.co.uk/news/science-environment-45775309

17 www.sustainabledevelopment.un.org/?menu=1300

18 www.dw.com/en/haiti-and-the-dominican-republic-one-island-two-worlds/a-16593022

Chapter 2

1 www.channel4.com/news/factcheck/no-plastic-bag-sales-arent-down-90

2 thegoodshoppingguide.com/blog/success-a-look-at-englands-new-plastic-bag-tax

3 medium.com/social-innovation-japan/7-surprising-facts-about-plastic-in-japan-f6920cc8e621

4 Ellen MacArthur Foundation www.ellenmacarthurfoundation.org/assets/downloads/EllenMacArthurFoundation_TheNewPlasticsEconomy_Pages.pdf

5 thediplomat.com/2018/04/indonesias-citarum-the-worlds-most-polluted-river/

6 www.plasticseurope.org/en/about-plastics/what-are-plastics/how-plastics-are-made

7 www.ecowatch.com/fossil-fuels-single-use-plastics-2565595371.html

8 www.express.co.uk/life-style/top10facts/948862/National-Tea-Day-tea-facts

9 www.nestandglow.com/healthy-food/plastic-foods

10 wedocs.unep.org/bitstream/handle/20.500.11822/25496/singleUsePlastic_sustainability.pdf?isAllowed=y&sequence=1

11 mainichi.jp/english/articles/20170819/p2a/00m/0na/002000c

12 news.nationalgeographic.com/2017/07/plastic-produced-recycling-waste-ocean-trash-debris-environment/

13 *How to Give Up Plastic: A Guide to Changing the World One Plastic Bottle at a Time*, Will McCallum, Penguin Life (2018)

14 www.ellenmacarthurfoundation.org/our-work/activities/new-plastics-economy/vision

Chapter 3

1 www.fashionrevolution.org/usa-blog/um-and-ah-the-psychology-of-self-justification/

2 truecostmovie.com/learn-more/environmental-impact/

3 www.forbes.com/sites/jamesconca/2015/12/03/making-climate-change-fashionable-the-garment-industry-takes-on-global-warming/#6921327e79e4

4 www.forbes.com/sites/jamesconca/2015/12/03/making-climate-change-fashionable-the-garment-industry-takes-on-global-warming/#734be43279e4

5 bettercotton.org/about-bci/cottons-water-footprint-how-one-T-shirt-makes-a-huge-impact-on-the-environment/

6 www.1millionwomen.com.au/blog/5-crazy-facts-new-fashion-documentary-true-cost/

7 'Are You a Consumer of Fast Fashion?', Trusted Clothes. Retrieved 24 April 2019 www.trustedclothes.com/Problem.shtml

8 truecostmovie.com/christina-dean-interview; ww.mckinsey.com/~/media/McKinsey/Industries/Retail/Our%20Insights/The%20State%20of%20Fashion%202019%20A%20year%20of%20awakening/The-State-of-Fashion-2019-final.ashx

9 www.contrado.co.uk/blog/what-is-polyester-a-closer-look-into-this-love-it-or-hate-it-fabric/

10 www.forbes.com/sites/jamesconca/2015/12/03/making-climate-change-fashionable-the-garment-industry-takes-on-global-warming/#6921327e79e4

11 truecostmovie.com/learn-more/environmental-impact/

Chapter 4

1 Podcast Deliciously Ella with Olio founder Tessa Clarke 'Reducing Food Waste'. deliciouslyella.com/podcast/food-waste-and-climate-change/

2 theconversation.com/five-ways-the-meat-on-your-plate-is-killing-the-planet-76128

3 foodrevolution.org/blog/food-and-climate-change/

4 climateandcapitalism.com/2018/06/26/why-avoiding-meat-and-dairy-wont-save-the-planet/u

5 sustainablefoodtrust.org/articles/sustainable-food-trust-welcomes-ipcc-report-which-recognises-positive-role-of-livestock-in-climate-change-mitigation/

6 www.theguardian.com/commentisfree/2018/aug/25/veganism-intensively-farmed-meat-dairy-soya-maize

7 www.ciwf.org.uk/media/7436369/how-to-transition-to-a-nourishing-sustainable-equitable-and-humane-food-system-2019.pdf10

8 www.daylesford.com/blog/daylesford-organic-chicken-sustainable-approach-feed-growing-appetite/

9 www.theguardian.com/lifeandstyle/2017/feb/14/sea-to-plate-plastic-got-into-fish

10 www.timetochangewales.org.uk/en/about/news/depression-and-suicide-are-leading-causes-death-within-farming-communities-uk/

11 blog.farmguide.in/farmer-suicides-in-india-suicide-notes-of-vishal-pawar-addc9d32183

12 www.theguardian.com/lifeandstyle/2005/aug/14/foodanddrink.features10

13 www.theguardian.com/global-development/2018/aug/20/food-waste-alarming-rise-will-see-66-tonnes-thrown-away-every-second

14 www.washingtonpost.com/news/theworldpost/wp/2018/07/31/food-waste/?utm_term=.515abce43b05

Chapter 5

1 trashisfortossers.com/3-ingredient-homemade-nontoxic-cleaner/

2 www.treehugger.com/htgg/how-to-go-green-laundry.html

3 www.theguardian.com/environment/2017/may/07/the-eco-guide-to-laundry

4 munchies.vice.com/en_uk/article/9kpjed/69-percent-of-brits-dont-know-food-waste-contributes-to-climate-change-survey-

finds?fbclid=IwAR2A6ykTGV75km8mEZP3G7kF2n83ROwFZvPj9pfpznaPRfY2zncOz9bf330

5 uk.whogivesacrap.org/pages/about-us

6 friendsoftheearth.uk/climate-change/renewable-energy-uk-how-wind-wave-and-sun-will-power-uk

7 www.linkedin.com/pulse/can-uk-run-renewables-alone-sian-smith/

8 www.theguardian.com/environment/2011/sep/16/carbon-offset-projects-carbon-emissions

9 matteroftrust.org/how-energy-efficient-homes-impact-the-environment/

10 www.atag.org/facts-figures.html

11 www.atag.org/facts-figures.html

12 www.atag.org/facts-figures.html

13 www.ontheluce.com/carbon-offsetting-flights/

Chapter 6

1 www.telegraph.co.uk/news/2018/08/01/decade-smartphones-now-spend-entire-day-every-week-online/

2 Boyd & Ellison, 2007, oxfordre.com/climatescience/view/10.1093/acrefore/9780190228620.001.0001/acrefore-9780190228620-e-369

3 oxfordre.com/climatescience/view/10.1093/acrefore/9780190228620.001.0001/acrefore-9780190228620-e-369

4 www.theguardian.com/commentisfree/2019/apr/15/rebellion-prevent-ecological-apocalypse-civil-disobedience?utm_term=Autofeed&CMP=fb_gu&utm_medium=Social&utm_source=Facebook&fbclid=IwAR2Qy_1AuKlhsye9N4LVdUAyO9sUOCz3vorPV6ObGbqf9VR9JnC41kxWPrM#Echobox=1555753873

5 www.nationalgeographic.com/environment/2004/01/consumerism-earth-suffers/

6 www.counterpunch.org/2016/12/20/rapacious-consumerism-and-climate-change/

7 www.robertsbridgegroup.com/our-views/business-be-warned

8 fashionista.com/2018/08/b-corp-certification-requirements-benefits-companies

Chapter 7

1 *Feminism and Ecology*, Mary Mellor, New York University Press,1997, p.1

2 www.undp.org/content/dam/undp/library/crisis%20prevention/disaster/7Disaster%20Risk%20Reduction%20-%20Gender.pdf

3 www.bridesofthesun.com/

4 www.unenvironment.org/news-and-stories/story/empowering-women-frontlines-climate-change

5 womendeliver.org/investment/invest-women-tackle-climate-change-conserve-environment/

6 www.actionaid.org.uk/about-us/what-we-do/womens-economic-empowerment/food-hunger-and-sustainable-livelihoods/how-climate-change-is-threatening-lives?gclid=EAIaIQobChMI_JO-8uu44wIVS7DtCh0c4gAMEAAYAyAAEgLAS_D_BwE

7 birthstrike.tumblr.com/

8 www.huffingtonpost.co.uk/entry/plastic-free-tampons-pads_n_5c0e88a6e4b06484c9fce988?guccounter=1&guce_referrer=aHR0cHM6Ly93d3cuZ29vZ2xlLmNvbS8&guce_referrer_sig=AQAAANIbzQOzBZ_jQtuMfVNdudKL8MboEThYnq7v6KWPH-fDgZ8yVfOs_UDfXL6ElxNBC4PUhxHVXuWaAzCdS7SFSwvj1pXhGxC2EkQLtt2Nye62W9E0IAsxt6bYarA8-0F1S3TkQP8sw5HQYdJV2oZ_kkFkVZwg5G6Y2QUu8u3xifky

9 friendsoftheearth.uk/plastics/plastic-periods-menstrual-products-and-plastic-pollution

10 zerowasteeurope.eu/zw-library/reports/

11 www.telegraph.co.uk/news/2018/11/13/wet-wipes-sold-flushable-responsible-93-percent-blockages-uk/

12 science.sciencemag.org/content/359/6377/760

13 www.theguardian.com/commentisfree/2019/apr/15/rebellion-prevent-ecological-apocalypse-civil-disobedience?utm_term=Autofeed&CMP=fb_gu&utm_medium=Social&utm_source=Facebook&fbclid=IwAR2Qy_1AuKIhsye9N4LVdUAyO9sUOCz3vorPV6ObGbqf9VR9JnC41kxWPrM#Echobox=1555753873

14 www.monash.edu/environmental-sustainability/news-and-events /latest-news/climate-change-a-first-hand-experience

15 ejfoundation.org/what-we-do/climate/ protecting-climate-refugees

16 medium.com/insurge-intelligence/escaping-extinction-through-paradigm-shift-83e33d4cb548